# The Ageless Generation

# The Ageless Generation

## HOW ADVANCES IN BIOMEDICINE WILL TRANSFORM THE GLOBAL ECONOMY

Alex Zhavoronkov

palgrave
macmillan

First published in 2013 by PALGRAVE MACMILLAN® in the U.S.—a division of St. Martin's Press LLC, 175 Fifth Avenue, New York, NY 10010.

Where this book is distributed in the UK, Europe and the rest of the world, this is by Palgrave Macmillan, a division of Macmillan Publishers Limited, registered in England, company number 785998, of Houndmills, Basingstoke, Hampshire RG21 6XS.

Palgrave Macmillan is the global academic imprint of the above companies and has companies and representatives throughout the world.

Palgrave® and Macmillan® are registered trademarks in the United States, the United Kingdom, Europe and other countries.

ISBN: 978-0-230-34220-0

Library of Congress Cataloging-in-Publication Data

Zhavoronkov, Alex.
   The ageless generation : how advances in biomedicine will transform the global economy / Alex Zhavoronkov
      pages   cm
   ISBN 978-0-230-34220-0 (hardback)
   1. Medical innovations—Social aspects. 2. Longevity—Forecasting. 3. Aging—Prevention. I. Title.
RA418.5.M4Z53   2013
613.2—dc23

                                                                    2012045840

A catalogue record of the book is available from the British Library.

Design by Letra Libre, Inc.

First edition: July 2013

10   9   8   7   6   5   4   3   2   1

Printed in the United States of America.

*To my parents, my grandparents.*

*To the many seniors, who continue to work, thrive, and continuously improve way past their retirement age.*

*And to those who dedicated their lives to science and medicine so we could live longer and healthier for the years to come.*

# Contents

# List of Figures

# Acknowledgments

Looking back at the time it took to write this book, it will be impossible for me to thank everyone who contributed. Many brilliant scientists and incredible people from all paths of life helped me, and it would take another book to be able to thank everyone personally.

This book would not have been possible without the help of Grady Cash, CFA. A retired U.S. Air Force pilot, futurist, and a writer himself, Cash could be a model for the lifelong learning and career-planning strategy for many baby boomers. I am grateful for his help with structuring, editing, and polishing the book. Another editor I would like to thank is Dr. Steven C. Scheer, a retired professor with a PhD in literature from Johns Hopkins, who is still actively working part time after surviving cancer, open-heart surgery, and health-related bankruptcy.

Karen Wolny, the strikingly intelligent editorial director at Palgrave Macmillan, reshaped the manuscript and provided an outstanding level of support and guidance. And it was my agent, Lynne Rabinoff, who immediately recognized the importance of the book and provided invaluable advice.

Simon Geletta, PhD, associate professor at Des Moines University, provided helpful research support with population trends and health-care policy. Valuable quantitative and qualitative research for the book was conducted by Jacek J. Slowikowski, a graduate of Columbia's Mailman School of Public Health. Both Jacek and Dr. Geletta worked with me for over nine months providing input and advice. Joseph Tarsio, PhD, MBA, also provided advice and performed a cover-to-cover edit, as did and Ross B. Finesmith, MD.

I would like to thank Nawazish Mirza, PhD, associate professor at LSE and a collaborator on other projects and papers, who helped with the demographic and economic forecasting. Igor Artuhov, PhD, contributed to the overview of the emerging technologies and the potential impact on demographic trends.

A long-time friend and mentor, Dr. Charles Cantor, a distinguished academic and the former director of the Human Genome Project at the Department of Energy, provided valuable advice. Dr. Cantor is truly a role model for healthy aging and continuous lifestyle improvement, and at 70 years of age spends most of his time traveling the globe starting and supporting research projects, developing novel diagnostic and therapeutic technologies, and giving lectures while maintaining an incredible level of fitness. And I would like to thank Elizabeth Cantor-Graae, PhD, one of the world's top experts on the epidemiology of psychologic disorders, and Edouard Debonneuil, PhD, the head of research and development (R&D) of AXA Global Life in France, for reviewing the book and making important edits. Augustinus Bader, MD, PhD, the head of Cell Techniques and Applied Stem Cell Biology department at Universitat Leipzig, and a long-time friend and colleague, offered significant insights on the future of cell and tissue regeneration and actively involved me in many regenerative medicine events and initiatives.

My good friend Aubrey de Grey, PhD, editor-in-chief of *Rejuvenation Research*, and his associate Michael Rae provided critical edits to the parts of the book explaining rejuvenation technologies. The ideas of William Millard, the founder of Computerland and the driving force behind the Lifestar Foundation, inspired the chapters of the book dealing with the propagation of the life-extending technologies into the government health-care systems. I would also like to thank my friend and collaborator Dr. Alexey Moskalev for letting me use his charts of the aging processes and advising on DNA damage and repair mechanisms.

The brilliant lawyers and economists Geoffrey Furlonger and Hugh Gallagher assisted me in understanding the current pension and benefits environments in Europe and the United States. Finally, this project benefits greatly from the illustrations and charts created by designer Adolfo Arranz.

# The Ageless Generation

# Introduction

Have you ever wondered, "Why do we get old?" Have you ever dreamed of being young again—regaining the energy and vitality you had as a 20-year-old? Have you ever wondered how you could make the world a better place for your children and grandchildren? I have. So has virtually everyone over the age of 30. This book was written for you.

We now stand on the brink of discovering answers to one of the greatest questions to face humanity: why do we age? We have begun to crack the code of life with breakthroughs in genetics and regenerative medicine that allow us to grow new organs and tissue. What was unthinkable a decade ago is now firmly in the realm of possibility. We will soon be able to slow the aging process itself. As far-fetched as it may sound, scientists have already done this in laboratory animals that are 99 percent genetically similar to humans. Even though mice appear to be so different from us, they have most of our genes and, like humans, develop cancer and diabetes and age in a strikingly similar ways.[1] New genetic therapies have allowed mice to live 28 percent longer—the human equivalent of a typical person living to age 105.[2] And that's just the *average* increase. A new stem cell therapy has more than *doubled* the life expectancy of a particular type of mouse—the human equivalent of over the age of 160.[3] It is only a matter of time—and research funding—until the benefits of these and other breakthroughs can be transferred to humans.

And this can't happen soon enough. Seniors aren't just getting older, they are getting *old*. The current medical system keeps seniors alive in spite of fatal illnesses, but not necessarily in good health. Instead, aging seniors survive one previously fatal illness only to contract another illness and then another. Today, per capita spending on senior health care is increasing faster than inflation, the number of seniors is increasing rapidly, and the ratio of workers to seniors is steadily decreasing. Advances in treatment of cardiovascular diseases and cancer are making a significant impact on mortality, but these new treatments are expensive. As a result, senior entitlement programs, such as Social Security and Medicare, are financially unsustainable at present levels. If spending on these programs is allowed to increase unchecked, the resulting global financial crisis could be far greater than anything previously seen in history.

When the consequences of a problem become too horrendous to contemplate, humans compensate psychologically by going into denial, acting as if the problem doesn't exist or clinging to a false hope that a proverbial white knight will ride in to save the day. This denial is clearly present in the dilemma of senior entitlement programs, where politicians have avoided these thorny issues for decades by passing the problems on to the next administration. Eventually, the situation will deteriorate to a point where our leaders will be forced to do something. But can they prevent a major economic crisis? When this question was posed to a think tank of 200 financial professionals, they responded with deafening silence.

Historically, a continued failure to address national overspending has led to dire results: hyperinflation, extreme unemployment, civil unrest, and ironically, as economies collapse, a loss of funding for the same senior entitlement programs that created the crisis in the first place. Poor financial management was one of the main contributing factors behind the advent of Nazi Germany as well as the collapse of the USSR that put millions of its senior citizens into poverty. As Aldous Huxley warned, "That men do not learn very much from the lessons of history is the most important of all the lessons that history has to teach."[4]

Will we learn from history? Fortunately, we may be bailed out by technology, which has advanced so rapidly in the past decade that a *medical* solution to these *economic* problems is now tantalizingly close. Through scientific means, we can dramatically enhance the health and youthfulness of the aging population over the next couple of decades. This would redefine our current conception of 65 as the standard age of retirement. If tomorrow's 65-year-olds were as healthy as 55-year-olds today, seniors could work an extra ten years if they chose to do so. If millions of seniors continued to pay into the system while postponing their entrance into these senior entitlement programs by a decade or more, the problems of Social Security and Medicare could be pushed decades into the future. And the cycle would continue, as medical researchers would have more time to extend even further the health and vitality of seniors, virtually eliminating the retirement age. As you will soon learn, the longevity breakthroughs we could see in the next 20 years could change the entire landscape of aging, including its social and economic implications.

To accomplish these lofty goals, policy makers, scientists, and funding organizations must work together to accelerate anti-aging and regenerative research while raising awareness that many age-related diseases and aging in general could soon be a thing of the past. It's not just about living longer; it's about living longer in good health without the myriad of physical conditions that slow today's seniors.

Billions of people will reach age 65 in the next 20 years. Many will approach these traditional retirement years with certain expectations defined by the existing concept of retirement, but their future will likely be far different. Many new and exciting discoveries for conquering the maladies of aging will be added to the arsenal of mainstream medicine in the coming years. Inevitably, these new technologies will lead to further refinements and even more advances. As a result, the world finds itself on the brink of an aging Renaissance, but can this new age of increased senior health and productive lifespans be achieved in time to save nations from the crushing burden of senior entitlement programs? Your future and the futures of your children and grandchildren could be riding on the answer. Let's begin by looking at what created this problem from a historical perspective.

# PART ONE

# The Era of Longer Lifespans

# ONE

# Approaching the
# Tipping Point

Decades ago, compassionate nations created welfare programs for aging seniors to provide a safety net against poverty in old age. At the time, no one anticipated that medical advances would dramatically increase life expectancy in the second half of the twentieth century, significantly increasing the size and cost of these programs. Today, senior programs are the fastest-growing and largest budget items for most developed nations.

Old-age pensions and senior health-care programs, such as Social Security and Medicare in the United States, are in desperate need of reform, but the challenges are almost insurmountable. Politicians can't even agree on where to start. Some advocate decreasing spending on these programs. Others, equally adamant, call for increasing revenue, a euphemism for increasing taxes. Still others advocate a combination of these approaches. Lost in this often-divisive rhetoric are some sobering facts. The number of seniors is increasing far faster than the number of youths entering the workforce. That means in the future there will be fewer workers paying taxes into Medicare and Social Security to support each senior. Today, the cost of these two programs is roughly $25,000 per senior.[1] That means that if we disregard everything else in the budget—education, national defense, transportation—the average worker would need to pay over $8,000 in taxes each year *to pay for these programs alone.*[2] As the number of workers per senior declines in the future, those costs will be even higher. The cost of these programs is so far out of line with reality that traditional solutions can do no more than temporarily postpone the inevitable financial crisis.

Fortunately, there is a workable solution. As remarkable as it may seem to those unaware of recent advances in anti-aging research, in the not-too-distant future medical science will possess the technology to slow and even reverse the aging process itself. Although this has gone relatively unnoticed by the general public, the past two decades have seen more advances in biomedical research than in the entire history of medicine. These advances are mostly occurring under the radar because they are often too complex to be explained in the typical 60-second news spot. They also might be years away from clinical application, which pushes them out of the major news cycle. Nonetheless, these advances are occurring at a breakneck pace.

The system I helped develop, the International Aging Research Portfolio, is one of the largest databases of government grant abstracts in the world. It tracks approximately 20 years of research funding by the National Institutes of Health (NIH), National Science Foundation (NSF), and almost a decade of research funding by the European Commission, Australia, and Canada. Although the database is far from complete, it shows that cumulative funding over the past 20 years exceeds half a trillion dollars. When private sector spending is included, total spending on medical research in the past 20 years probably exceeds $1 trillion.

We are now beginning to see substantial dividends from this research. Life expectancies are topping age 80 in many nations. Past research is also paying dividends in the sense that it laid the groundwork for faster and more frequent advances in the future. Just ten years ago, the only way to do stem cell research was to destroy embryos. Since 2008, new, promising types of stem cells have been created in laboratories that offer significant medical and ethical advantages over embryonic stem cells because they can be created without using embryos at all. National initiatives in regenerative medicine—those focused on developing medical technology to repair age-related tissue damage, grow new organs, and restore the lost function—could help alleviate the looming crisis of old-age pensions and senior health care.

Unfortunately, it will take a long time for the possibility of extreme longevity to reach mainstream acceptance. When I casually remarked to a young medical student that she and many of her peers might live beyond 100, she laughed at what seemed to be an outrageous idea. Yet in spite of such widespread skepticism, the concept of extreme longevity will eventually reach what Malcolm Gladwell calls "the tipping point" in his book of the same name. A major objective of this book is to help bring about that tipping point much sooner than it would otherwise occur.

Fortunately, regenerative medicine is in a far more advanced state than most people realize. Scientists have increased the life span of *C. elegans*—a type of worm—by ten times. Fruit flies—another common laboratory test subject—have lived four times longer than normal. Genetic therapies have allowed mice to reach the equivalent age of 160 in human years. This is particularly significant because mice are so genetically similar to humans. Hearts have been grown from a single cell and successfully transplanted into living, breathing animals. Humans have achieved a functional age that is 15 years younger than their biological age. Cancers have been cured in animals that are genetically very similar to humans. The pieces of the technological and medical puzzles to extend longevity and, more to the point, *healthy* longevity, are now coming together. The remaining pieces, or at least enough pieces to make a dramatic change in the health of seniors, can be found within a decade—*if there is sufficient research funding to make it happen.* The basic technology already exists to pursue research in a number of areas, all of which show promise to dramatically increase health span and life expectancy in tomorrow's seniors. What is lacking is a national sense of urgency and a strategic plan—a roadmap, if you will—to achieve these goals. In the 1960s, the United States set a seemingly impossible task, to put a man on the moon, and achieved that goal in less than ten years. We need a similar commitment today for aging research initiatives.

Another risk is that if the United States doesn't act soon, it might not be able to retain the new jobs associated with this research within its borders, relinquishing the subsequent economic boost to other nations. In a rise similar to the emergence of Japan as a technological superpower in the 1970s, China and India are rapidly catching up with the United States and Europe in biomedical research. In 2010, China revealed plans to invest $1.5 trillion over the next five years on high-tech industries, including biotechnology. At the 2011 International Conference for Bioeconomy, China announced plans to invest $308.5 billion in biotechnology research over the next five years. China expects this research to generate 1 million jobs, extend life expectancy by one year, and reduce childhood mortality to 12 percent of current levels.

In 2009, Novartis, a major pharmaceutical company, announced it would invest $1 billion in research and development in China over the next five years, including a significant expansion of its Biotech Medical Research Center in Shanghai. Those investments and jobs are now lost to the United States, but the loss of future investments and jobs could be prevented by a new national policy to create a more favorable environment for medical research and regenerative medicine.

## THE STRUCTURE OF SCIENTIFIC REVOLUTIONS

The pace of technology is accelerating so rapidly that it's hard to keep up—worse, it's hard to change old ideas. Beliefs that have existed for years or even throughout one's lifetime are very hard to erase, no matter how convincing the science is behind them. In effect, a lag exists between the time a new discovery changes the existing scientific reality and the time it takes for scientists and the general public to accept that new reality as fact.

In theory, this obstacle shouldn't exist. Albert Einstein once said, "No amount of experimentation can ever prove me right; a single experiment can prove me wrong."[3] The simple elegance of Einstein's statement captures the purity of science. One solid experiment—properly documented and independently replicated—should be all that is necessary to change scientific opinion, but in a world filled with scientific and human bias, this isn't quite what happens.

In his landmark book, *The Structure of Scientific Revolutions*, Harvard scientist Thomas Kuhn proposed that science does not move forward solely on objective criteria or by the linear accumulation of new knowledge. Instead, scientific "fact" is defined by the current consensus of the scientific community. When new discoveries are made, scientists are reluctant to accept anything that conflicts with widely held scientific beliefs. According to Kuhn, scientific advances must first be confirmed by enough scientists until the new theory reaches a tipping point of acceptance. At this point, it replaces the old scientific theory. Kuhn coined the phrase "paradigm shifts" to indicate the tipping point at which the consensus shifts from one major belief structure to another.[4]

Unfortunately for both medical researchers and for individuals suffering from age-related conditions, this shift of scientific consensus can lag behind the discoveries by years. The history of medical science is rife with examples of dogmatic beliefs slowing the acceptance of medical advances. It took the medical profession over half a century to accept the benefits of hand washing as fact, the merits of which Ignaz Semmelweis discovered in 1848. Max Planck, winner of the Nobel Prize in 1918 for his discoveries in quantum physics, once dryly observed, "A new scientific truth does not triumph by convincing its opponents . . . but rather because its opponents eventually die." Later, Planck would simply quip, "Science advances one funeral at a time."[5]

Today, the medical consensus is that aging cannot be slowed or stopped, but it is only a matter of time until that paradigm shifts. Current medical advances are already keeping seniors alive longer, but not necessarily in good health. Whereas seniors survive one illness that would have been fatal only a couple of decades ago, they only live long enough to contract another equally expensive disease. The cost to senior healthcare programs will be staggering.

With so much at stake, why are only a few visionary scientists calling for more research designed to combat aging and restore function? Why has it been so difficult for this exciting new paradigm to capture the imaginations of scientists worldwide? As Kuhn explains, change takes time, yet the pace of change is now accelerating so rapidly that neither world leaders nor many scientists are able to keep up with it.[6]

In previous generations, medical research was a painstakingly slow process, but today, it's a different ballgame thanks to the personal computer. Moore's Law—first defined by Gordon Moore in the late 1960s—forecasts that computing power will double every two years.[7] This law has proven to be amazingly accurate in predicting the increase in computing power that has in turn exponentially increased the pace of medical discoveries. While skeptics believe that Moore's Law will only continue for a few more years because of the physical limits to how small a silicon computer chip can be made, engineers are poised to overcome the limitations with the next generation of computers—optical, DNA, spintronics, and even chemical computers that will mimic the neurons in the human brain. By the time silicon chips reach the point where they cannot be further miniaturized, one of these new computer technologies is likely to emerge in the computer of the future. The implications of the continued doubling of computing power every couple of years are almost unimaginable. Projects that took years to complete at a cost of millions of dollars will be within the reach of high school students for their science projects.

Consider, for example, one of the greatest computer projects ever undertaken. In 1989, the National Institutes of Health established the National Human Genome Research Institute with the stated purpose of sequencing the human genome. This project would take a decade to complete, and its results would eventually determine the sequence of about 3 billion chemical base pairs that make up DNA, while also identifying the 23,000 genes found in humans. About ten years ago, when this project was completed, the 13-year cost of sequencing a human genome for the first time was $3 billion—in large part due to the immense amount of computing power needed for the task.[8]

Today, thanks to rapidly increasing computer power, genome sequencing can be done for under $10,000, and the cost continues to come down. For just $99, you can obtain a comprehensive test of tens of thousands of SNPs (single nucleotide polymorphisms or DNA sequence variations) from 23andMe, a company, named after the 23 pairs of chromosomes in human DNA, founded by Anne Wojcicki, the wife of Sergey Brin, the cofounder of Google. The process is simple. Consumers send a saliva sample to 23andMe by mail and then later access the results online. I took this test in 2009 and periodically receive updates with new information about my DNA. More than 125,000 people now use 23andMe and other similar services to find out more about their ancestors, predisposition to several diseases, and carrier status for a variety of genes associated with disease. As databases grow, this process could eventually identify new ways for treating and diagnosing diseases, advance diagnostic medicine, and create a database that links genetic profiles with increased risk of specific diseases.

Another reason that medical technology is advancing so rapidly is that advances in many nonmedical fields are now contributing to medical science. Take mathematics, for example. You may not have heard of the Fourier transform (FT)—a mathematical technique that simplifies information—but it is used to compress vast amounts of information in everything from JPEG files to MP3 music. In medicine, FT takes the data collected during MRIs and turns it into images that assist physicians in diagnosing illnesses. Mathematical algorithms are used ubiquitously in the medical field in applications ranging from data analysis to diagnosis. Mathematics is also used in a multitude of computer models, such as using fluid dynamic equations to understand blood flow through veins and arteries. Medical images, such as MRIs and ultrasounds, are also created using mathematical formulas.

Some medical advances draw on technology so advanced that it may seem right out of *Star Trek*. Nanotechnology is bringing together the fields of mechanical engineering, information technology (IT), and mathematics to create tiny machines from individual atoms that would be only a few nanometers wide—thinner than a single strand of human hair. Over the years, the field of nanotechnology has expanded to include not just machines and computers but many other exciting applications. In the future, nanotubes embedded in clothing could change color at the wearer's whim or switch from breathable fabric to waterproof as the need arises. Military scientists envision nanotube fabrics that harden upon impact, creating an impenetrable barrier to projectiles, while otherwise retaining the comfort characteristics of regular fatigues. Nanotechnology shows enormous promise for medicine as well.

In 2006, scientists at Stanford University reported that near-infrared lasers—a wavelength slightly higher than visible light and harmless to human tissue—could heat man-made carbon nanotubes to 158°F in two minutes.[9] These nanotubes were then injected into a cancerous tumor and, after allowing sufficient time for them to be absorbed by the cancer cells, were exposed to near-infrared light. The cancerous cells were killed, while adjacent cells remained unharmed.

Since thousands of nanotubes can fit inside one cell, exposing them to near-infrared lasers effectively allows scientists to target cancer at the cellular level. An obstacle to this approach had been the difficulty of delivering the nanotubes only to cancer cells and not to nearby healthy cells, but this was overcome by coating the nanotubes with folate, a B vitamin. Unlike normal healthy cells, cancer cells contain numerous folate receptors, making them essentially a magnet to folate-treated nanotubes.

This method is still in the experimental stage, but if it proves successful, near-infrared lasers could be the cancer cure for which scientists have long dreamed. The tools are also relatively inexpensive. Low-power infrared lasers are already in widespread use as noncontact digital baby thermometers. If nanotechnology-delivered cancer treatments survive clinical trials, this procedure could transform medicine. Today, cancer is the number-two killer of seniors, second only to heart disease, but the near-infrared laser, among many other advances in biomedical sciences, promises to cure cancer with a short series of inexpensive, noninvasive treatments with no side effects.[10]

In some cases, exciting new technologies become obsolete before they even reach widespread availability. One such device is the proton therapy machine. Proton beam therapy is an extremely sophisticated, computer-controlled technique that treats cancer with a thin beam of protons with an accuracy of less than 1 mm, or the width of the tip of a pencil lead. This extreme localization minimizes the damage to surrounding tissue often experienced in traditional radiation therapy.

Unfortunately, each machine requires a cyclotron, which accelerates protons to sufficient speed to be effective. Cyclotrons are circular devices as big as a football field and cost more than $100 million to build. As a result, there are only 13 such machines in the United States, so there is a long waiting list for patients to utilize proton beam therapy.

But the future is bright for cancer patients, as new technologies such as near-infrared laser therapy promise to minimize damage during treatment at far less cost. Promising advances that attack cancer at the individual cell level could replace proton beam therapy before most people even learn that proton therapy exists.

## COMMUNICATING AT THE SPEED OF LIGHT

Medical breakthroughs are seldom the result of one great person with one brilliant idea. Instead, they are the result of a collection of previous discoveries passed on from one researcher to the next. For example, although Louis Pasteur is widely credited with developing the germ theory of disease, it was actually built upon the discoveries and observations of other scientists. The successful development of penicillin by Howard Florey would not have been possible had the previous penicillin discoveries of Alexander Fleming not been passed on to him. Fleming himself based his research on the successes of sulfa drugs that had been communicated to him through colleagues and medical papers.

Prior to the advent of computers, it was difficult to obtain research information that might eventually contribute to major medical advances. The discovery of penicillin was arguably the greatest medical development of the past 100 years, but it took nearly ten years from the time Alexander Fleming published his landmark paper until Ernst Chain, a biochemist at Oxford University, discovered Fleming's paper while looking through piles of research articles on antibiotics. Ironically, Oxford University was less than 60 miles away from Fleming's laboratory in London. Sixty miles in ten years is an effective delivery speed of 84 feet per hour. To call this a snail's pace would be an insult to the common garden snail, which has been clocked at 158 feet per hour. That difference might seem trivial, but prior to the widespread use of penicillin, more soldiers died from infections than battlefield trauma. During World War I, for example, 1.1 soldiers died from infections for every battlefield trauma. By the end of World War II, the ratio had dropped from 1.1 to 0.06.[11] If penicillin had been widely available in 1940 instead of 1945, think of how many lives could have been saved.

Until recently, finding medical research information required traveling to the nearest medical library and poring through stacks of journals in hopes of finding an article on the desired subject. Now, medical students in China can immediately access research in London—or anywhere else in the world for that matter—from their dorm rooms. Millions of research papers can be screened for keywords, prioritized, and presented for orderly review in a matter of seconds.

The ability of scientists worldwide to share information today is unprecedented in human history. The rise of the internet has dramatically increased the ability of researchers to share information and benefit from the findings of other researchers. In 1995, the National Library of Medicine, which is a part of the NIH, launched a global tool called PubMed, an online database where all notable scientific journals submit publication abstracts. By June 1997, PubMed searches reached 2 million per month. By 2006, searches topped 3 million per day.

The easier and faster scientists can communicate, the faster medical advances can occur. Scientists around the world can now communicate almost instantaneously. It's not unusual for me to collaborate with associates in four continents in one day from my apartment in Moscow or Los

Angeles. Similar communications are taking place in medical research every day.

## MEDICAL BREAKTHROUGHS ACCELERATE

Only a few decades ago, it was rare to see more than one major medical breakthrough each year. Today, medical breakthroughs occur almost every week. In 2010 alone, the world saw several medical advances that would have been major news a few decades earlier. An artificial pancreas successfully completed clinical trials. This device maintains blood sugar within acceptable levels using an insulin pump connected to a computerized sensor. It allows diabetic patients to lead a normal life without the blood sugar spikes that eventually cause long-term damage.

For the first time, a patient with a mechanical heart was deemed sufficiently stable to go home from the hospital while awaiting a donor heart. Researchers from the Queensland Institute of Medical Research discovered genetic variants that increased the risk of ovarian cancer, which could eventually lead to earlier detection of this disease. Scientists in Australia and America discovered a way to treat cancer using a new class of genes called microRNAs—molecules that regulate gene expression. The biotechnology company Geron received FDA approval to begin testing an embryonic stem cell treatment for spinal cord injuries—the world's first stem cell testing on humans.

And the list just keeps growing. The world's first full face transplant was performed. Scientists at Johns Hopkins University developed a simplified method to turn blood cells into beating heart cells. Another team of scientists from Johns Hopkins discovered a way to create a new type of stem cells from human hair follicles. Scientists at the University College London created functioning heart muscle in mice using another type of cell present in the hearts of adult subjects. This could pave the way to regenerate human heart tissue using cells from the individual's own heart, bypassing the ethical and immunological challenges of using embryonic stem cells. Cells taken from the tips of mouse tails were genetically reprogrammed and used to repair damaged liver tissue using a relatively new process called transdifferentiation.

As remarkable as these advances are, even more amazing break-throughs await. By the time a child born today reaches the age of 65, medical knowledge will have increased so dramatically that if everything currently known about medicine could be placed inside a book the size of this one, in 65 years the newly discovered medical knowledge would fill hundreds or even thousands of such books. Eventually, these discoveries will address ways to slow or even reverse some aspects of aging.

The increasing pace of technology is clearly a driving force in medical breakthroughs and the recent rapid growth of the senior population, as shown in Figure 1.1. The line with the sharp upslope shows how popu-lation has increased over the past several centuries, while the comments show technological advances. Beginning in 1400, it took 400 years for the population to double from 500 million to 1 billion, and another 150 years to double again, but then it took just 50 years to more than triple to 7 billion. Technological progress shows a similar pattern. This is not to imply that the population will continue to grow at tremendous rates. Birthrates are trending down in most developed nations. Instead it shows that technology drives longevity, both through better living conditions and medical advances. In the lower right-hand corner of this chart, a sharp upward sloping line shows the unprecedented rise in the senior population over the past 50 years.

Until just recently, changes from generation to generation through-out recorded history have been quite slow. Often, there was very little to differentiate one generation from the next—or one century from the next, for that matter. With the exceptions of the rise and fall of ancient Greece and Rome, each successive generation from Classical Antiquity to the Middle Ages would experience basically the same level or degree of civilization as the generation before and after it.

Then, about 200 years ago, changes in civilization began to increase—slowly at first, then gradually accelerating to the point that now each new generation is experiencing a world that is dramatically different from that

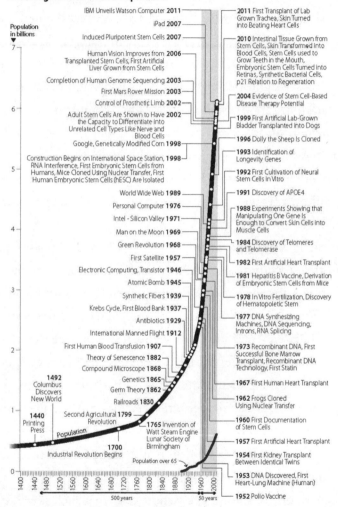

*Figure 1.1.* World population, population over 65, and technological progress.

of the previous generation. Much of the breakthroughs in science and technology that led to the advent of the digital age and globalization came to light during the past half century. The most notable technological advances, such as portable electronics and the internet, were widely adopted in the past two decades—in just one generation's time. Research and clinical practice is now communicated and coordinated via online papers, emails, and forums at an unprecedented speed. As a result, there has been considerable progress in the biomedical sciences with advances in biomaterials; diagnostics methods; cell, tissue, and organ engineering and regeneration; genomics; proteomics; and bioinformatics.

The road from discovery to medical application is unfortunately much more difficult for biomedical technologies because advances require years of clinical testing. Consequently, many recent advances that could significantly extend healthy life spans are still bogged down in the red tape surrounding medical research.

Fortunately, some life-extending technologies are already in or nearing mainstream clinical use. Statins, beta-blockers, vasodilators, and other pharmaceuticals that significantly extend the survival of patients with cardiovascular conditions became blockbuster drugs with tens of billions of dollars in revenue. The mainstream clinical use of many of these drugs and procedures started less than three decades ago and focused on treating the patient. The focus on patients is a natural starting point to make health technologies marketable. But the greater effect is likely to come when those technologies will be used to prevent the development of the medical condition in the first place, before symptoms become troublesome.

At present, due to the relatively short time that such novel drugs and treatments have been available and because they target the relatively small subset of the population with serious health conditions, their long-term effects on life expectancy cannot yet be estimated. Consequently, it may take another decade before we begin to see a significant impact on life expectancy from these advances, but breakthroughs are coming faster each year. Many life-extending technologies are currently being developed in

R&D centers around the world. Strong pressure from aging populations is likely to provide more incentive to reduce the time until these experimental procedures reach clinical use.

There is no doubt that the boundaries of aging will eventually be pushed to the point where 65 will no longer be considered old. These advances can't come too soon for nations struggling under the burden of debts incurred by senior entitlement programs, such as Social Security and Medicare. Technology alone cannot guarantee these advances will come in time to prevent serious economic hardship from these unsustainable expenses. In spite of the urgency of this coming economic crisis, some promising technologies are thwarted by bureaucratic red tape or languish due to a lack of funding. Meanwhile, other research projects with dubious medical breakthrough potential receive ongoing government grants.

How did we get here? How did we get to the point where these humanitarian programs now pose such a tremendous financial burden for nations around the world?

# TWO

# The History of Longevity

As nations struggle to pay for old-age pensions and senior health-care programs, increasing life expectancies are making these problems even worse. Younger readers might wonder what their elected officials were thinking when they created these monstrous cash-hungry programs that now threaten the financial security of entire nations. It's easy to forget that until the second half of the twentieth century, the likelihood of living past age 65 was very rare.

It's also easy to forget that until recently, the concept of retirement didn't exist. Modern retirement—15 or more years of voluntary disengagement from the workforce to enjoy travel and leisure activities—is an even newer concept, having only received widespread acceptance beginning in the 1970s. Prior to that, retirement was a relatively brief period of limited activity, poor health, and frugal living. In fact, when the concept of retirement was first introduced about 100 years ago, life expectancy was only age 47.[1]

We know—thanks to archaeologists' analysis of fossilized bones—that prehistoric man had a life expectancy of about 22 years. That remained relatively constant for tens of thousands of years during which time *Homo sapiens* evolved, as would any other species, to optimize its survival through those very arduous and often dangerous 22 years. Throughout the millennia, our distant ancestors struggled to survive harsh climates, hunger, and dangerous animals. Only the fittest survived long enough to achieve reproductive success and pass on their genetic makeup to the next generation and generations to follow, thereby increasing their genetic advantages in the overall population of *Homo sapiens.* The priorities for survival of the species were quite different in those days. There was no old age as we think of it today because the older and weaker members of the population usually died from something other than natural causes—hunger, wild animals, drought, famine, accidents, or weather extremes—long before they could reach their optimal life span, let alone the maximum possible age.

## LIFE EXPECTANCY VERSUS LIFE SPAN

The terms "life expectancy" and "life span" are often used interchangeably, but their meanings are considerably different, which can lead to confusion. "Life expectancy" is an actuarial term that indicates the age at which exactly half of a given population still survives. As this is written, life expectancy for a newborn in the United States is 78.3 years, which means that out of every 100 babies born this year, 50 will live to age 78.3. Of course, some—half in fact—will live longer. In order to avoid confusion, "life span," for the purposes of this book, will be defined as the maximum possible age that can be achieved by a member of a particular species. For humans, maximum life span today is 122, the age achieved by Jeanne Calment of France in 1997.

Understanding the difference between life expectancy and life span helps to explain certain discrepancies in historical archives that would otherwise be quite puzzling. In 399 BCE, life expectancy was only 25, but the ancient Greek philosopher Socrates had already reached age 70 when he was famously sentenced to death by drinking hemlock. Hippocrates, who lived during the same period and is widely acknowledged as the father of medicine, was even more sturdy, living to the ripe old age of 90. Although they died well short of the maximum human life span of 122, both of these ancient Greeks lived far past the average life expectancy at the time.

## FROM PREHISTORIC MAN TO ANCIENT GREECE

In geological terms, *Homo sapiens* is a relative newcomer to this planet, having evolved from the genus *Homo* in Africa about 200,000 years ago. It would take another 150,000 years before modern *Homo sapiens* developed. In this case, "modern" is an anthropological term, meaning these early humans exhibited cultural characteristics similar to those of modern man, such as burial rituals and the ability to make tools, artwork, jewelry, and music.[2]

While 50,000 years is a mere blink of the eye in geological terms, from an evolutionary standpoint this would represent roughly 3,000 generations of reproduction and evolutionary advances. Yet in all of those generations, life expectancy remained relatively unchanged. From the arrival of modern *Homo sapiens* 50,000 years ago until the rise of ancient Greece 2,000 years ago, life expectancy increased to only 25 years of age. Over the next two millennia, life expectancy would slowly rise, reaching age 35 by the time of American Independence in 1776. However, that increase was more the result of improved living conditions than medical advances.

Of course, these early estimates of life expectancy are at best educated guesses since accurate written government records of births and deaths have only been kept for just over 100 years. Estimating the maximum life span of early humans is even more difficult. Since living conditions were so harsh, we can reasonably assume that the major limiting factor was not biological, but rather the inability of humans to survive to old age in the harsh conditions of the time.

Still, living beyond age 35, while rare thousands of years ago, was certainly not unheard of. Hieroglyphic records and autopsies of mummies show that many Egyptian pharaohs lived into their 40s or 50s. According to the ancient Egyptian historian Manetho, the first pharaoh of the first Egyptian Dynasty, Menes, who ruled Egypt around 3000 BCE, lived to age 62.[3] It's unlikely that most people have ever heard of Menes, but you probably know him by his other title, the Scorpion King—immortalized in cinema by the movie *The Mummy Returns.* Hieroglyphs found at a burial site believed to be that of Menes included a symbol for a scorpion and a victory over another proto-dynastic ruler. However, records of his rule are sketchy. Some historians believe that the Scorpion King was an earlier pharaoh or possibly Hor-Aha, Menes' successor as pharaoh.[4]

While the life span of the Scorpion King is debatable, multiple records exist for Djer—Hor-Aha's son—who reigned for over 41 years, so we know that Hor-Aha lived to at least age 41. By the reign of Rameses II from 1279–1213 BCE, more complete historical records are available to us. Rameses II ruled for a remarkable 66 years and died at either age 90 or 91.[5] Unlike the previous examples based solely on historical records, scientists have confirmed his unusually long life from autopsies performed on his mummy, which is now on display in the Cairo Museum. Interestingly, although Rameses II suffered from crippling arthritis and probably walked with a hunched back in his later years, the most likely cause of his death stemmed from complications brought on by an abscessed tooth.[6]

Obviously, fossilized remains of elderly humans would be extremely rare since so few lived to advanced age, but several specimens have been found that were more than 35 years old at death. Archaeologists in Spain have discovered what might have been an ancient burial site over 530,000 years old that contained the remains of several specimens of *Homo heidelbergensis* (an ancestor of *Homo erectus*) that were older than 35. One fossilized pelvis bone found at this site indicated an age at death of over 45.[7]

But these examples were the exception. For 200,000 years, humankind lived through roughly 12,000 generations of evolution with no change in its primary directives: be hardy enough to reach sexual maturity under very harsh conditions, reproduce, and live a few more years until offspring could survive independently. Only in the last 100 years, or roughly five generations, have these evolutionary imperatives changed. Interestingly, the groundwork for this unprecedented increase was laid over 2,400 years ago.

## A BRIEF HISTORY OF MEDICINE—
## STARTING WITH HIPPOCRATES

Prior to 400 BCE, most scholars believed illness was punishment from wrathful gods. The Greek doctor Hippocrates correctly theorized that illness resulted from physical causes. By separating it from religion, medicine could henceforth be pursued by the scientific method of cause and effect rather than the prevailing religious dogma.

Despite Hippocrates' groundbreaking approach to illness and the resulting paradigm shift in thought about disease, the scientific discoveries to advance his beliefs would not occur for more than 2,000 years. In 400 BCE, infectious disease was the number-one killer and would remain so until the middle of the twentieth century. Poor sanitation and hygiene spread disease from person to person. Child mortality was alarmingly high—some historians estimate that over 30 percent of infants died before the age of five. Many people died from misguided medical practices that had no basis in science. Finally in the nineteenth century, physicians made several major discoveries that would contribute to the increased longevity we enjoy today.

## HYGIENE AND ANTIBIOTICS

In the 1800s, it was common practice for doctors to go straight from the autopsy room to seeing patients. The inevitable spread of virulent infections was often catastrophic, especially for women in childbirth. The death rate for pregnant women from "childbed fever" was five times higher in hospitals than for mothers who delivered babies at home.

In 1847, Hungarian physician Ignaz Semmelweis realized childbed fever was contagious and could be drastically reduced if physicians washed their hands before moving from one patient to another. The risk was especially high if the caregiver had been handling corpses prior to treating pregnant women. When Semmelweis implemented a regimen that required all caregivers to wash their hands before seeing each patient, the mortality rate dropped tenfold within three months, clearly showing that the spread of disease could be dramatically reduced by hand washing.[8]

Armed with seemingly unequivocal data, Semmelweis lectured on his discoveries in 1850, but the medical community, which still considered bloody physician clothing as a sign of professional success, responded with scorn. Semmelweis became obsessed with his beliefs. A decade later in 1861, he would publish a book on his theories, but it would also be met with derision from the medical community. He continued to fight for the acceptance of hand washing, but in 1865, he suffered a nervous breakdown. He was committed to an insane asylum where he died shortly thereafter.

About a decade after Semmelweis' discovery, Louis Pasteur, a young French chemist, discovered that disease-causing germs existed everywhere and could be spread to patients by doctors.[9] Pasteur's discovery of what is now called the germ theory of disease validated Semmelweis' theories, but Pasteur was also met with the same intransigence from the medical community.

In spite of the proven efficacy of hand washing, it would be resisted by medical experts for decades. In 1879, at the Academy of Medicine in Paris, Pasteur himself would argue with a speaker who cast doubt on the benefits of hand washing.[10] You might think that such narrow-mindedness would be ancient history, but even today, the medical establishment remains stubbornly locked in the old ways of doing certain things. In spite of a growing body of evidence to the contrary, many otherwise intelligent and informed physicians reject the premise that aging can eventually be slowed. Many promising research projects languish behind walls of medical dogma and bureaucratic red tape, as modern day Semmelweises and Pasteurs meet similar resistance to new ideas.

In the 2,400 years from ancient Greece to 1900, the primary limiting factor on life expectancy would continue to be infectious diseases, which would still account for over one-third of all deaths. Deadly outbreaks of cholera, dysentery, tuberculosis, typhoid fever, influenza, yellow fever, and malaria were commonplace. The flu pandemic of 1918 killed tens of millions of people. Over 450,000 deaths were recorded in the United States, mostly among otherwise healthy people under the age of 40. In India, 16 million died. Another infectious disease—tuberculosis—was responsible for one out of every six deaths in France in 1918.

As virulent as infectious diseases used to be, by the end of the twentieth century these diseases would be almost completely wiped out by three medical advances: public health initiatives, vaccines, and the greatest medical advance over the past 2,400 years: penicillin. Ironically, in spite of the billions being spent on medical research today, the most profound medical advance of the twentieth century—spurring the biggest increase in life expectancy in the history of humankind—was the result of a laboratory accident.

In 1928, Alexander Fleming, a 47-year-old Scottish biologist working at St. Mary's Hospital in London, was searching for a new antibacterial drug because existing drugs were ineffective. Before leaving on vacation, Fleming placed dirty Petri dishes in a basin filled with Lysol, a disinfecting agent that would allow the dishes to be reused. There were so many dishes that not all were submerged. Upon his return, Fleming noticed that a blue-green mold had grown on the dry dishes, killing the bacteria that once lived there. Probing further, Fleming discovered that the mold secreted a bacteria-killing substance. He named the substance penicillin and published his findings in 1929.[11]

Unfortunately, penicillin was very difficult to produce and could not be made in large enough quantities for clinical studies. No further progress was made until Ernst Chain, a biochemist at Oxford University 60 miles away, came across a paper authored by Fleming. Chain showed this paper to his supervisor Howard Florey, professor of pathology at Oxford, and requested permission to perform some experiments. Florey agreed but otherwise showed little interest until after Chain's first successful test on mice. Florey and Chain then conducted other successful experiments, but the inability to produce the drug in sufficient quantities still hampered their efforts. It took several months and a staff of 17 workers to produce a single dose.[12]

When England was swept into World War II, Florey and Chain were faced with an urgent and widespread need for antibacterial drugs. Since Florey was head of the department, he approached the Rockefeller Foundation and received funding to conduct penicillin research. He then searched the United States to find the best source for penicillin and found it in moldy cantaloupe in Peoria, Illinois. For mass production, corn was used since it was available in far larger quantities than cantaloupe.[13]

Even so, penicillin was still difficult to produce. It took two years from Florey's trip to the United States before mass production began. From January to May 1943, only 400 doses of penicillin had been made, barely enough to treat one shipload of sailors returning from local brothels after shore leave. By the end of the war, however, U.S. production of penicillin had increased to over 650,000 doses a month. Its success in treating infections after World War II led to penicillin being dubbed "the wonder drug."[14] In 1943, Florey and Fleming were knighted for their efforts. In 1945, Ernst Chain finally received the recognition he deserved when he, Florey, and Fleming won the Nobel Prize for medicine.[15]

After the use of penicillin became widespread, deaths due to infectious diseases decreased dramatically. Heart disease quickly became the number-one killer in America, followed by cancer. Between 1900 and 2000, cancer-related deaths increased over 300 percent. Today, heart disease and cancer account for 55 percent of all deaths in the United States. Meanwhile, death from infectious diseases has dropped from more than 35 percent of all deaths in 1900 to less than 5 percent today.

## LIFE EXPECTANCY BY COUNTRY

The differences in life expectancy by country are quite dramatic, ranging from just over 39 years in Mozambique and Swaziland to almost 83 years in Japan. Many factors affect life expectancy, including individual and state welfare, health care, education, behavior, and environmental factors. Studies have also demonstrated that there is a positive correlation between wealth and life expectancy. Wealthier and more educated people can afford better health care, more nutritious diet, and exercise. In the developed world, life expectancy has been steadily increasing for both men and women as demonstrated in Figure 2.1.

As Thomas Kirkwood, a famous biogerontologist—a scientist who studies what happens to the body as it ages—put it in his BBC Reith Lecture in 2001, "In the course of the last 50 years, life expectancy has continuously increased by 2 years every 10 years. That is equivalent of adding, each decade, about 20 percent more time to live."[16]

In theory, these figures could have been much greater if it was not for the negative changes in diet and behavior. In some of the developed countries, we can clearly see the longevity benefits stemming from the technological advances and increases in welfare being virtually erased by high-calorie diets; alcohol, tobacco, and drug use; and lack of exercise and physical activity. Russia is a dramatic example of how each of these factors contributes to this disconnect. Even though it is ranked fifty-fifth by the International Monetary Fund in income per capita, the life expectancy of

## Life Expectancy at Birth and at Age 65

| | Life expentancy at birth | | | | Life expentancy at age 65 (years) | | | |
|---|---|---|---|---|---|---|---|---|
| | Females | | Males | | Females | | Males | |
| | 1990 | 2008 | 1990 | 2008 | 1990 | 2008 | 1990 | 2008 |
| United States [1] | 78.8 | 80.3 | 71.8 | 75.3 | 18.9 | 19.8 | 15.1 | 17.1 |
| Australia | 80.1 | 83.7 | 73.9 | 79.2 | 19.0 | 21.6 | 15.2 | 18.6 |
| Austria | 79.0 | 83.3 | 72.3 | 77.8 | 18.1 | 21.1 | 14.4 | 17.7 |
| Belgium | 79.5 | (NA) | 72.7 | (NA) | 18.8 | (NA) | 14.3 | (NA) |
| Canada | 80.8 | (NA) | 74.4 | (NA) | 19.9 | (NA) | 15.7 | (NA) |
| Czech Republic | 75.5 | 80.5 | 67.6 | 74.1 | 15.3 | 18.8 | 11.7 | 15.3 |
| Denmark | 77.8 | 81.0 | 72.0 | 76.5 | 17.9 | 19.5 | 14.0 | 16.6 |
| Finland | 79.0 | 83.3 | 71.0 | 76.5 | 17.8 | 21.4 | 13.8 | 17.5 |
| France | 80.9 | 84.3 | 72.8 | 77.6 | 19.8 | (NA) | 15.5 | (NA) |
| Germany | 78.5 | 82.7 | 72.0 | 77.6 | 17.7 | 20.7 | 14.0 | 17.6 |
| Greece | 79.5 | 82.5 | 74.6 | 77.5 | 18.0 | 19.9 | 15.7 | 17.7 |

NA: Not Available. (1): Source of 2008 life expectancy data: US National Center for Health Statistics, National Vital Statistics Reports (NVSR), "United States Life Tables", Vol. 58, No. 21, June 2010, and unpublished data.
Source: Except as noted, Organization for Economic Cooperation and Development (OECD), 2011, "OECD Health Data", OECD Health Statistics database (copyright), www.oecd.org/health/healthdata, accessed April 19, 2011.

*Figure 2.1. Life expectancy by country at birth and at age 65.*

its citizens is a meager 70.3 years and ranks one hundred twelfth in the world.[17]

In the United States, high-calorie foods and sedentary lifestyles contribute to slowing down increases in life expectancy. According to the recent OECD (Organisation for Economic Co-operation and Development) report summarized in Figure 2.2, almost 34 percent of U.S. adults are obese.

The life expectancy of a baby born today in the United States stands at 78.3 years. While this is a dramatic increase from a century ago, the country doesn't even rank in the top ten in this category. The longest-lived nation is Japan with a life expectancy of 82.6 years followed by Hong Kong at 82.2 years. According to the United Nations, the United

## Percentage of the Adult Population Considered Obese
2008 data

| Country | Percentage |
|---|---|
| United States | 33.8 |
| Mexico | 30.0 (2) |
| New Zeland | 26.5 (1) |
| Australia | 24.8 (1) |
| United Kingdom | 24.5 |
| Canada | 24.2 |
| Ireland | 23.0 (1) |
| Luxemburg | 20.0 (1) |
| Hungary | 18.8 (4) |
| Greece | 18.1 |
| Czech Republic | 17.1 |
| Finland | 15.7 |
| Spain | 14.9 (2) |
| Belgium | 13.8 |
| Germany | 13.6 (3) |
| Austria | 12.4 (2) |
| Denmark | 11.4 (3) |
| France | 11.2 |
| Norway | 10.0 |
| Sweden | 10.0 |
| Italy | 9.9 |
| Switzerland | 8.1 |
| South Korea | 3.8 |
| Japan | 3.4 |

(1): 2007 data. (2): 2006 data. (3): 2005 data. (4): 2003 data.
Source: Except as noted, Organisation for Economic Co-operation and Development (OECD), 2011, "OECD Health Data", OECD Health Statistics database (copyright), accessed April 2011. See also www.oecd.org

*Figure 2.2. Obese population by country.*

States ranks a distant thirty-sixth in life expectancy, which is below Cuba and Costa Rica.[18]

## THE NEW DISEASES OF OLD AGE

As death from infectious diseases decreased thanks to penicillin and antibiotics, diseases that take longer to develop moved to the forefront. Prior to the advances that controlled infectious diseases, illnesses like cancer and heart disease were uncommon because these diseases often take years or even decades to develop. It's the same for the other major killers of seniors, such as diabetes and Alzheimer's, which were virtually nonexistent. Today, medicine must face this new reality: most people are living into, and dying at, a far more advanced age.

## LIFE EXPECTANCY AT AGE 65

Earlier, we defined life expectancy as the age to which exactly half of a given population would survive. For a newborn child in 1950, life expectancy was age 67. However, for adults who beat the odds and survived through early-stage diseases and accidents, life expectancy, once they reached age 65, was 78.9.

In the first half of the twentieth century, life expectancy at birth increased by more than 20 years, primarily because of higher survival rates in infancy, childhood, and early adulthood. But with the exception of infectious diseases like pneumonia, influenza, and blood poisoning, very little progress was made in treating diseases of the elderly. In fact, for those fortunate enough to survive the diseases and early childhood killers in the 1800s, life expectancy *at age 65* was about the same in 1850 as it was in 1950. This would begin to change dramatically in the second half of the twentieth century with advances in chemotherapy for cancer and improved surgical procedures for heart and cardiovascular diseases.

Beginning in 1950, life expectancy at age 65 began to increase by about 0.8 years per decade and continued to do so for the next 40 years.

Not coincidentally, many medical advances occurred during this period that dramatically improved seniors' chances of surviving previously fatal illnesses. In the 1950s, the first experimental heart surgeries were performed, and by the late 1960s, heart bypass surgeries were available. By the 1990s, doctors began using less invasive surgical procedures, allowing procedures like bypass surgery to be performed on seniors in their 80s. Today, thousands of octogenarians in the United States undergo major heart surgery each year. This would have been unthinkable three decades earlier, when the invasive surgery itself would have most likely been fatal.

Suddenly, beginning in 2000 in the United States, life expectancy at age 65 began to increase faster, jumping from 0.8 years per decade over the past 50 years to 1.0 years over the next seven years—a rate of 1.5 years per decade![19] Although there was no readily apparent reason, the most likely cause was the widespread availability of improved surgical techniques and cancer treatments for seniors in the 1980s and 1990s. As demonstrated in Figure 2.3, cancer survival increased steadily over the years due to better diagnosis and treatment.

As a result, these elderly seniors, having survived previously fatal illnesses, now live past their earlier life expectancies in large numbers. There's little reason to suspect that this trend will slow down in the

## Increasing Cancer Survival Rates

Relative survival (percent) by the year of diagnosis (all cancer sites, all races, males and females) in the United States 1975-2008

| Survival | 1975-79 | 1990 | 2000 | 2005 | 2006 | 2007 | 2008 |
|----------|---------|------|------|------|------|------|------|
| 1-year   | 69.9    | 75.6 | 79.5 | 80.7 | 81.3 | 81.6 | 81.8 |
| 2-year   | 60.1    | 67.1 | 73   | 74.4 | 75.4 | 75.7 |      |
| 3-year   | 55      | 62.7 | 69.7 | 71.2 | 72.2 |      |      |
| 4-year   | 51.6    | 59.8 | 67.6 | 69.2 |      |      |      |
| 5-year   | 49.1    | 57.9 | 66.1 |      |      |      |      |
| 6-year   | 47.2    | 56.2 | 64.9 |      |      |      |      |
| 7-year   | 45.5    | 54.7 | 63.8 |      |      |      |      |
| 8-year   | 44.2    | 53.5 | 62.8 |      |      |      |      |
| 9-year   | 43      | 52.4 | 62   |      |      |      |      |
| 10-year  | 41.9    | 51.4 |      |      |      |      |      |

Compiled from the National Cancer Institute's Surveillance, Epidemiology and End Results Cancer Statistics Review, 1975-2009.

*Figure 2.3. Increases in cancer survival rates.*

future. In fact, it's likely that it will accelerate in the coming years due to the remarkable medical advances discussed later.

## THE RISE OF RETIREMENT

Today, many workers consider retirement to be a right, as if it were something that has existed throughout history. But they might be surprised to learn that it is a relatively new concept. When George Washington completed his second term as president of the United States at age 65, he didn't retire. Instead he returned to Mount Vernon, Virginia, where he worked for the rest of his life as a farmer and whiskey distiller. On December 12, 1799, at the age of 67, Washington inspected his fields, spending several hours on horseback in freezing rain and snow. The next day, he awoke with a severe sore throat that some historians believe was a complication of tonsillitis. Washington died the following day.[20]

Even in the first half of the twentieth century, retirement was virtually unknown except among those who were physically or mentally disabled and unable to work. Thomas Edison—inventor of the incandescent light bulb and America's most prolific inventor with more than 1,000 U.S. patents—was typical of that era's workers. Edison certainly had the resources to retire had he wished to do so. Instead, at age 65, he managed a factory complex that covered more than 20 acres and employed 10,000 people. So not only is retirement a relatively new concept, but it was actively resisted by older workers, who viewed it as being "put out to pasture"—a euphemism for sitting around and just waiting to die.

Some people mistakenly believe the advent of retirement was the result of the Social Security Act of 1935, but the concept was born much earlier as a byproduct of the changing workplace. The seeds of retirement were first sown after the Civil War in the United States with the rise of the Industrial Revolution. In agriculture, the work was hard, but speed wasn't essential. An older farmer could start working a little earlier or work a little later to make up for any age-related loss of physical strength. However, in the first half of the twentieth century, the rise of the industrial age required workers to put in long hours on assembly lines. That type of work was not well-suited to seniors because of the accelerated pace of work. Assembly lines could only operate as fast as the slowest worker. Older workers, who were no longer as strong or as fast as younger workers, slowed the entire production line.

As the economy became increasingly driven by industry instead of agriculture, employers needed a way to get rid of older, less productive workers so they could replace them with younger, cheaper ones. Many employers instituted mandatory retirement age. The era of the "gold watch" was born: "Please accept this gold watch as a token of our appreciation for your 30 years of service—now go away." Since life expectancy in those days was so short, some companies hit on the idea of giving these new retirees a small stipend—a pension for their old age. Although this was a first step, it still wasn't retirement as we know it today. These early stipends were seldom enough to support the senior for very long, and as a result, most workers viewed retirement with hostility or a sense of fatalism.

With the arrival of the Great Depression, many older retirees stopped getting even these meager pensions. They were penniless and unable to find jobs. Thus, the Social Security Act of 1935 was implemented as the government's response to the rise in abject poverty of retirees with no pensions or savings, but it didn't actually *create* retirement.

Even after the passage of the Social Security Act, resistance to mandatory retirement continued for decades. From a senior's perspective, retiring was identified not by what it was but what it was not—not working, not being productive, and not having a meaningful role in life. It meant removal from the work force, which throughout seniors' adult lives had given them their identity and sense of purpose. In effect, retirement was defined in those days as being useless, waiting for the inevitable loss of independence, and then death. The work ethic prevalent at the time wasn't consistent with leisure activities—not that there were many leisure activities to begin with. Because retirement was still a relatively new concept, most retirees' friends and neighbors were still working, so seniors were further segregated from their community, leading to a greater sense of isolation. It's no wonder these early retirees found themselves floundering in a limbo of negativity.

The first major attempt to reshape modern retirement into what it has become today began in 1954 when Ben Schleifer, a realtor in Phoenix, Arizona, opened a retirement village with the Pollyanna-ish name of Youngtown.[21] The village never prospered, but it caught the eye of construction magnate Del E. Webb. Webb grew up in turn-of-the-century California, with ambitions of becoming a professional baseball player, but at age 16, he was left near death after a severe bout of typhoid fever. After recovering, he moved to Arizona for his health and began working in construction. By World War II, Webb had established himself as one of Arizona's largest contractors. He bought a share of the New York Yankees in 1945 and worked with Howard Hughes to build several plants for Hughes Aircraft Company. Webb constructed apartments, hospitals, hotels, and military bases, and once boasted that he had met every president from Franklin Delano Roosevelt to Richard Nixon.[22]

In 1960 at the age of 61, he founded the Del E. Webb Corporation with a public stock offering of $12 million, nearly $100 million in today's dollars. Today, Webb is the largest developer of retirement communities in the United States—managing over 50 retirement communities in 20 states.[23] His focus on retirement communities began in 1954 when some of his executives heard about Schleifer's Youngtown project on a local television station. They decided that Schleifer's modest development, which had targeted retirees with very limited incomes, was not the best approach, so they chose to build a far different type of retirement community.[24]

Armed with deep pockets to fund a far more ambitious undertaking, Webb built the nation's largest planned community exclusively for seniors. It featured a recreation center, golf course, swimming pool, and shopping center—all of which was completed before the first home was sold. A massive pre-opening advertising campaign assured prospective buyers that their new community would provide not only the health benefits of living in Arizona, but a new way of life for seniors—a concept that advertisers called "active retirement." On opening day, January 1, 1960, the road into the new community was backed up for two miles as 100,000 people flocked to see this glimpse of the future. In the first three days, 250 homes were sold.[25] By the end of the month, an additional 150 homes were under contract to be built.[26]

This new town—appropriately named Sun City—would go on to become the largest and best-known retirement community in the United States. In a stroke of marketing genius, Sun City's "new way of life" advertising campaign did what decades of time and hundreds of employers had failed to do. Del Webb, in the space of only a few years, redefined retirement as a positive stage of life. Sun City's brochures promised that residents would enjoy the best years of their life there, finding greater fulfillment, health, and longer life—all available at very affordable prices. As a nod to the prevailing work ethic of the day, their advertising described retirement as something seniors had "earned" through their years of labor. According to one of their advertisements promoting this new way of life, living in Sun City got retirees off the sidelines and back into the game of life.

The opening of Sun City was only part of Del Webb's influence on the new concept of retirement. After its blazing fast start, interest in Sun City cooled from over 1,500 home sales in 1960 to only 400 sales per year in 1963. Fortunately, Webb had surveyed its existing home buyers and found that residents included an exceptionally large number of professional and white-collar retirees with relatively high retirement incomes. Most of these buyers had eschewed conventional financing and paid for their homes outright. Many had also extensively remodeled their homes at substantial personal expense.[27]

After gaining a better understanding of Sun City's demographics, Webb chose to ignore the low-income market and instead focused his marketing on more affluent retirees. More home choices were offered. Extra amenities—gates, courtyards, garages, and walk-in closets—were added. Telephone and power lines were laid underground to reduce the visual clutter, further increasing the image of an upscale community. A larger and more lavishly equipped recreation center was built adjacent to an attractive man-made lake. As a result, the price of the average home increased from $15,540 in 1965 to $25,000 in 1969. In spite of the rise in price, Webb had correctly identified the market and his company sold more homes than ever. Leisure activities were as abundant as their marketing campaign promised. Sun City residents were living the good life while their peers suffered through the harsh winters of the Midwest and other northern climes. In the space of less than a decade, Sun City had become the model for the retirement dream.

By 1972, Sun City's population had more than doubled over the previous four years, rising from 11,000 to 25,000. By 1978, it had quadrupled to 44,000. What originally began as a small senior village had become Arizona's eighth-largest city. The desirability of retirement communities—and thus the acceptance of retirement itself—began to skyrocket. Older workers—tired of shoveling snow—saw these Sun City advertisements and longed for their piece of the American dream. If they couldn't afford to retire in Sun City, they would retire somewhere else.[28]

Seeing the success of Del Webb's Sun City, entrepreneurs rushed to develop similar communities in Florida. In order to conserve costs, most of these communities were built using mobile homes, rather than site-built housing. As a result, these communities were so affordable that retirees could keep their northern homes, visiting family and relatives up north in the summer while escaping to Florida to avoid the harsh winters. "Snowbirds," as these part-time residents were called, would become so omnipresent that they would more than double the population of many small central Florida communities each winter.

While one could argue that the change in perception of retirement was gradual, looking back, we can tell that this cultural change was relatively abrupt when it finally occurred—taking place primarily from 1960 to the mid-1970s. Prior to 1960, seniors were ambivalent or openly hostile toward retirement. By 1978, retirement as we know it had not just become accepted; it had become part of the American dream.

## RISE OF DEMOCRACY

For most of the previous millennium, nations were ruled by monarchs, dictators, or feudal warlords. Some rulers were benevolent, but even so there was no incentive for individuals to excel in business because their profits and even their lives could be forfeit at the whim of totalitarian rulers. One of the first major nations to embrace democracy was the United States, adopting its Constitution in 1787 with France quickly following in 1789.

With the ascension of democracy, worker conditions improved and so did life expectancy. Entrepreneurs rose to prominence for the first time in human history. Industrial and banking barons like Cornelius Vanderbilt, John D. Rockefeller, Dale Carnegie, J. P. Morgan, Baron Rothschild, and Henry Ford forged empires for themselves and employed tens of thousands of people. In the early twentieth century, individual rights were further protected by the creation of unions and collective bargaining.

Armed with individual rights and fair pay for a fair day's work, workers began to buy items previously reserved for aristocrats and kings. The production of consumer goods soared to keep pace. Washing machines, air conditioners, televisions, and automobiles were mass produced, creating even more jobs at unprecedented rates as national economies grew faster than at any previous time in recorded history. To keep these goods affordable and available to the masses, employers moved factories overseas, globalizing production in a movement that continues today.

Social programs, relatively rare in the early twentieth century, surged under democracies as citizens, imagining themselves in similar old-age situations in the future and considering Social Security as the fundamental right for their hard work, voted for elected officials who supported legislation to create senior health care and pension safety nets. Most governments established old-age pensions and health-care programs for their senior populations in the first half of the twentieth century. Upon signing Medicare into law in 1965, the United States became one of the last developed nations in the world to create a health-care program for its seniors.

When Social Security—established in 1935—paid its first monthly checks to beneficiaries in 1940, the ratio of workers to seniors was 16 to 1. Today, it has dropped to only 2.9 to 1 and the ratio is continuing to decline as large numbers of baby boomers join the swelling ranks of the senior population.[29] The dwindling ratio of workers to seniors has already begun to create significant financial problems for senior entitlement programs around the world.

## THE RISE OF GLOBALIZATION

Further exacerbating the declining ratio of workers to seniors was the rise of globalization—the trade of goods and services among countries. Of course, trade among distant nations has existed for centuries. Marco Polo famously established trade routes between Europe, Central Asia, and China in the thirteenth century, but commerce was limited to a narrow range of nonperishable products that were both light and valuable enough to be transported vast distances at a profit. As a result, the majority of commerce was local or regional until the machines of the rapidly expanding industrial age made trading with distant populations possible. By 1950, oranges from Florida and peaches from South Carolina could be shipped to New York and still be affordable to middle-income families. The diesel-electric train and diesel-powered ships made it possible for corporations to locate their new factories in other countries thousands of miles away where the cost of labor was much less. They could then ship their products back to their home country at such a low cost per unit that these products could be sold for more profit than if they had been manufactured in a local factory.

The growth of markets from local to regional to global had a profound effect on both developed nations and developing countries. In the United States and the European Union, corporations thrived due to the lower cost of producing goods in developing nations. Consumers benefited because products cost less. The developing countries also benefited—despite cases of worker exploitation—because their economies grew much faster than before. Factories located in these countries brought jobs, stability, and structure to developing communities. While the employers often paid very low wages, these wages were still higher than the workers could find anywhere else. Since the employer had a vested interest in worker productivity, employees often received better health care and food than they would have received in the absence of these factories. As a result, the economies and standards of living in many of these developing nations are currently growing faster than those of the United States, Europe, Japan, Russia, India, and China.

On the downside, the relocation of these factories offshore resulted in a loss of jobs in the United States and Europe. At first, the rapid growth of the national economies offset these job losses, so the loss of jobs was only a concern to the displaced employees. But as the populations aged and overall economic growth slowed, this trend toward industrial relocation led to higher unemployment, which in turn made it more difficult to generate sufficient tax revenue to sustain senior welfare programs such as Medicare and Social Security.

Prior to the Industrial Revolution, these programs were less necessary because seniors could continue to work in some capacity on the farm to an advanced age. Today, with the rise of the Information Age, we've come full circle in the sense that there are no significant reasons why the majority of the information-age jobs couldn't be handled by seniors well into their 70s or even 80s. In fact, seniors are better suited for some information-age jobs than younger workers because they have more emotional maturity and contextual experience.

## THE RISE OF SENIOR HEALTH CARE

As life expectancy increased, of course, retired seniors now faced many diseases that had been rare in the past, notably cancer and heart disease.

Since few seniors on a fixed income had the resources to pay for treatment, most developed nations created subsidized health-care programs. In most countries, it's a form of universal health care. In other nations, it's typically a combination of government and private insurance plus contributions by the employee and employer.

Unfortunately, the cost of these well-intentioned programs has spiraled out of control. In 1970, Medicare expenses averaged $300 annually per recipient. By 2010, these expenses had skyrocketed to $12,000 per beneficiary, an increase of over 13 percent annually. Inflation, of course, impacts the cost of health care, but it alone would have only increased the average payment to $1,417. Ironically, the majority of these cost increases come from medical advances. Consider the case of Margie Harris.

In 1998 at age 82, Margie was diagnosed with severe heart valve problems. A couple of decades earlier, she would have been told to rest and been given some medication. She would most likely have lived only a few more years. However, as result of advanced medical technology, Margie had heart valve replacement surgery and survived for another ten years. Unfortunately, her health began to decline almost immediately after the surgery. In the next few years, she would develop diabetes, Alzheimer's disease, and experience two strokes. She would eventually be admitted into a nursing home for the last four years of her life. The medical cost incurred by the family and Medicare jointly over those last ten years of life exceeded one-third of $1 million.[30]

Margie's situation was in no way uncommon. Consider Helen, an active 80-year-old widow in apparently good health when a routine checkup discovered a similar heart problem. The surgery was successful, but during convalescence, she fell and broke her hip. Thus began a spiral of heartbreaking events. She fell again and broke her shoulder. She developed pneumonia and suffered a stroke. Her mild diabetes became uncontrollable and she developed dementia. Helen was admitted into a nursing home for the last three years of her life. Although no one calculated the exact cost to Medicare, it was well in excess of $250,000.

Margie's and Helen's case studies illustrate the paradox of modern health care for seniors. Medical advances keep seniors alive in spite of illnesses that would have previously been fatal, but those seniors will almost certainly develop other illnesses, which in turn create even more Medicare expenditures. These expenses are further exacerbated by changing demographics. The world is inexorably turning gray.

# THREE

# Demographics of Major Economies

*The United States,
the European Union,
Russia, China, and Japan*

Populations around the world are not just aging, they are getting *old*. The medical advances of the twentieth century dramatically increased life expectancies the world over. The result has been a change in national demographics, cultures, politics, and the traditional family structure.

Families today are radically different from those of a hundred years ago. In the 1800s, families were large and close knit. It wasn't uncommon for couples to have six children or more because more children meant more hands to work on the farm in a predominantly agricultural society. As parents aged, they could undertake less demanding chores when their physical abilities began to decline, leaving the harder work to the younger generation. Grandparents who could no longer work lived under the same roof and were cared for by their adult children and grandchildren. Even after children reached adulthood, families stayed close because it was unusual for offspring to move far from their ancestral home. In those days, most people grew up, lived, and died within a few miles of where they were born—with the exception, of course, of the pioneers who settled the American West.

Today, families are much smaller, averaging 2.1 children in the United States, less than 1.8 in Europe, 1.6 in China, and only 1.3 in Japan.[1]

If birthrates continue to decline throughout most of the developed world, as most experts anticipate, even more financial problems for pensions and senior health care will arise. In the United States and the European Union, the ratio of workers to seniors has already declined to the point that these programs are no longer self-supporting by contributions from employees and employers.

If governments respond by cutting funding for these programs, the burden of care for seniors will fall increasingly on adult children. However, the rise of the industrial age required young adults emerging from high schools and universities to move away from home in search of

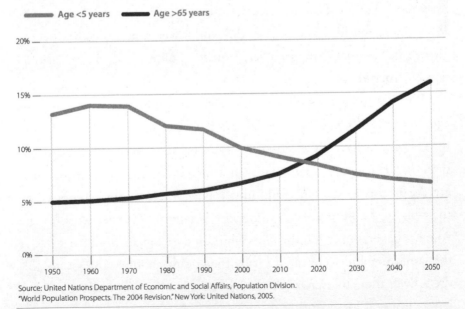

## Young Children and Older People as a Percentage of Global Population

Source: United Nations Department of Economic and Social Affairs, Population Division. "World Population Prospects. The 2004 Revision." New York: United Nations, 2005.

*Figure 3.1. Young children and older people as a percentage of global population.*

employment. Although family members still tend to live closer together in Europe where the countries are smaller, it's not uncommon in the United States for siblings to live thousands of miles apart. This makes it more difficult for adult children to provide care for aging parents.

Shirley Krieger, age 84, lives alone in Winter Haven, Florida. Her health has begun to decline, especially her mobility, and she's becoming more forgetful. A hundred years ago, she would likely have moved in with one of her four adult children who could look after her. But, in our highly mobile society, they all live in distant states. This increases her living expenses, of course, because she must hire a handyman and a housekeeper to perform lawn maintenance and some household chores, but it also increases the potential cost for Medicare. Without adult children nearby to urge her to see a physician, minor illnesses often progress to major problems before she sees a doctor. Total compliance with drug and medical treatment protocols occurs less often.

Once Shirley's health declines to the point she can no longer live by herself, the cost of assisted living would be borne by Medicaid. Multiply this situation by millions of seniors and it's easy to see how this will create substantial financial liabilities for senior health care in the future.

The situation is similar in China—the youth swarm to glistening new cities where older faces are noticeably absent. Meanwhile, in small rural villages throughout the country, the youth are disappearing, leaving their parents behind. This is a rising concern for China, because eventually someone will need to care for that aging population.

The geographical separation of the extended family worldwide will make it much more difficult for adult children to provide care for aging relatives in the future. There is also less cultural pressure to do so. A century ago, when everyone knew their neighbors and attended the same social gatherings, adult children experienced far more cultural pressure to provide support for their elderly parents. Today that cultural pressure has almost disappeared in Western nations, although it is still a factor in Asia.

In the past, the elderly could count on their families in times of ill health, but now, for someone with no spouse and adult children spread throughout the land, there are far fewer options available. The failure of large numbers of baby boomers to save sufficient funds for retirement is likely to make this situation even worse. At some point, poor health will force them to quit working, and with no family nearby, many will be forced into assisted living facilities. Once their meager savings have been depleted, Medicaid will be forced to cover those expenses, which means even more senior health-care expenses for states and the federal government.

## RISING SENIOR POLITICAL POWER

When Shirley and her husband retired 30 years ago to Winter Haven in central Florida, it was a sleepy, isolated town about an hour's drive from Orlando. Their new retirement community was surrounded by orange groves and giant water oaks dripping with Spanish moss, an idyllic life in their golden years—a realization of the American retirement dream. Shirley and her husband had traveled to Arizona to consider Del E. Webb's Sun City, but they finally decided on Florida because it was closer to their home state of Illinois and to the home of Shirley's older sister, who lived in nearby Lake Alfred.

When their children and grandchildren used to visit them in Florida, they would drive through the countryside and the orange groves to visit Disney World, less than an hour away. Today, the orange groves have disappeared, replaced by dozens of housing developments and thousands of homes dotting the countryside. Most of those homes are owned by seniors who, like the Kriegers, moved south in an attempt to escape the harsh winters of the Northern Tier states. As a result, Florida has become a demographically old state—the oldest state in the United States.[2]

Florida's older-than-average population has made it a bellwether for senior issues, especially during presidential election campaigns. Every four years, presidential hopefuls travel to Florida to tout their platforms and explain why they would be the best candidate for seniors. These candidates cannot afford to ignore Florida, which has more electoral votes than any state in the nation other than California and Texas, and as many as New York. Florida's political influence becomes even more evident when one considers that since 1972, Florida has voted for the winning candidate in 10 out of the last 11 presidential elections. It missed supporting the winning candidate only once—the landslide election of Bill Clinton over George H. W. Bush in 1992, when independent party candidate Ross Perot captured nearly 20 percent of Florida votes.[3]

In 2000, seniors composed 17.5 percent of the Florida population. Of course, those under the age of 18 can't vote, so seniors constituted 22.7 percent of potential voters. When only likely voters are counted—adults deemed most likely to cast a ballot in the next election—the percentage of senior voters is even higher. A smaller percentage of young people tend to vote than seniors, especially during midterm elections when far less than half of young people cast ballots.[4] Roughly seven out of every ten seniors, on the other hand, tend to vote in each presidential and midterm election—a little higher in some years, a little lower in others, depending on the importance of senior issues.[5] When considering likely voters instead of the voting age population, seniors cast roughly three out of every ten votes in each Florida election.

The road to the presidency travels through Florida, of that there can be no doubt, but when we consider how quickly other states are aging, it will soon be apparent that senior voters will have more power than ever in the future. By 2025 the percentage of seniors nationwide will top that of Florida in 2000. During every presidential election, political pundits opine over the importance of various red and blue states, but in the near future another political powerhouse will be added to the mix—the rise of gray states.

There are two important lessons we can take away from the rise of seniors in Florida. First, the senior voting block is extremely important. Politicians can ignore it only at great peril to their political futures. Second, Florida is a harbinger of things to come for the rest of the United States and for the rest of the world. That is, the rest of the world with the notable exception of Japan, where the age of seniors has already arrived.

## THE ELDER STATESMAN OF NATIONS—JAPAN

One in five Japanese are already over the age of 65. By itself, that would be no cause for alarm, but the number of seniors is poised to increase rapidly in the near future. What's more, Japan is about to experience something that is extremely rare among nations in modern history—a long-term decline in population.

Throughout history, populations have rarely declined, except during periods of plague, famine, or war. In part, that was due to high fertility rates—it was not uncommon for women to have several children—but today the birthrate is falling dramatically. This does not bode well for Japan, not only because a large percentage of Japan's population is already over age 65, but also because Japan's birthrate is among the lowest of industrialized nations. Meanwhile, as more Japanese baby boomers reach age 65, its elderly population is poised to skyrocket—doubling from one in five today to about two in five of Japan's total population by 2050.

Once children under working age are omitted from these population figures, the ratio of workers to seniors will be about one to one. In other words, the taxes on one Japanese worker would have to support the old age pension and health-care expenses of one senior. To put that enormous cost in perspective for readers in the United States, one U.S. worker would have to be taxed sufficiently to pay all the Social Security and Medicare benefits for one senior, which today have reached a combined cost of over $25,000 annually. Of course, that's only the expense for Social Security and Medicare. The government would need even more revenue to run other programs, such as defense and education, as well as pay the salaries and pension plans for government employees.

Traditionally, in Japan, seniors care for the very young and the younger adults care for the very old. These large extended families typically live under one roof—a relatively small roof by Western standards—but over the past few years, families have gotten smaller and young adults have moved to the big cities. Many older Japanese now believe that they will have to fend for themselves in old age rather than be cared for by their children. Still, old traditions die hard. Interviews with young Japanese—some sporting Western fashions and spiked hairstyles—find that many still believe that they will follow the traditional role of caring for aging parents in their old age.[6]

The Japanese aren't alone in this concern. Already many working parents in other nations—mostly women in their 40s and 50s—are simultaneously caring for their own young children as well as for their aging parents who have grown too frail to live by themselves. Demographers call this the sandwich generation—middle-aged parents who fall into the stressful role of dual caregivers.

For Japan, there seems to be no relief in sight. Young women are showing little interest in marriage, preferring to stay independent and working rather than having children. The anticipated result is a decrease in future workers that are part of an extended family lending support to Japan's increasingly elderly population. How Japan handles this crisis will be a lesson for democracies around the world, because similar trends are occurring in other developed nations.

## THE AGING DRAGON—CHINA

Perhaps in no other place on earth will the growth of the senior population be more dramatic than China. In 2009, over 120 million people were over the age of 65.[7] By 2050, that population is projected to almost triple—larger than the projected population of the entire United States.[8] By 2050, one in four Chinese is projected to be over 65.[9] Longevity is also poised to increase dramatically. Life expectancy is currently about 73, but it's much higher in large cities, where China is aggressively expanding to accommodate millions more workers.

Fortunately for China, it is in a unique position in terms of its size and economic clout and its unique blend of socialism and capitalism. China can mandate sweeping changes at a pace that would be impossible in a democracy. Just three decades ago, China implemented its "period of adjustment," which transformed its mostly agricultural society into the world's number-one exporter of goods. This economic miracle has been driven by a youthful workforce and aggressive government policies to grow the economy, but the supply of new workers has begun to slow down. The one-child policy implemented in 1979—limiting couples to only one child per family—has resulted in a scarcity of young workers that will soon create a massive aging problem in China. As more women join the workforce and postpone having children or never procreate at all, the replacement ratio—the number of children per woman—has already dropped below one in some of the major cities. Demographers cite 2.1 as the minimum ratio needed to keep a population from long-term decline.[10]

That means there will be far fewer young adults entering the workforce in the coming decades. Beginning in 2035, the population of China will peak and gradually begin to decline. Today, there are five workers for every senior, but by 2050 this ratio will dramatically decline to less than two workers per senior.[11] So far China has successfully compensated for the growing shortage of children by creating initiatives to ensure that a very high percentage of its young are in the workforce. China has made massive investments in education, for example, and it now produces more college-educated workers than any other nation in the world—more than the United States and India combined.[12] This has ensured that more women are in the workforce than ever, but increased education has had the undesirable side effect of women marrying later and having fewer children, which will further exacerbate China's aging problem in the future.

China has also made massive investments in infrastructure, building new cities at a dizzying rate. In order to keep its population happy, China has to rapidly grow the economy, which increases the quality of life for its citizens. But rapid growth also requires more jobs. China's goal is to provide an educated and motivated workforce and the infrastructure to make it easy for large multinational corporations to build their new factories in China.[13]

Culturally, China is caught between its ancient traditions and the new values of an increasingly urbanized society. Historically, adults toiled in the fields while the elders cared for the very young as long as they were able to do so. Then the extended family cared for the elders. Today, millions of young Chinese have fled to the cities for jobs, leaving behind only the very old and very young in the remote villages. Still, old cultural traditions die hard and are often modified to fit the new reality of modern life. Seniors are leaving their existing friends and family to follow their children into the cities, where they pick up children from school and do the shopping while parents are at work.

## EUROPEAN UNION

Most of Europe is facing a similar trend, although not quite as dramatically as China. Beginning around 2010, the working-age population is predicted to decline by 0.5 to 1 percent per year over the next 40 years.[14] That's roughly the same rate as China, but China has a much larger number of seniors who will be retiring, so the net decline in ratio of workers to seniors will be much greater in China than in most of the rest of the developed world.

Making long-term demographic projections in Europe is more difficult, however, because of immigration. The European Union has an estimated 20 million legal immigrants, but the number of illegal immigrants is estimated to be somewhere between 3 and 6 million. These immigrant populations have higher birthrates than the native populations, so the long-term effect of immigration could have a significant impact on the future ratio of workers to seniors. Assuming these immigrants assimilate into the native culture, it could alleviate the declining birthrates of these nations.

Immigrants can also have a significant cultural impact since Europe is less of a melting pot than the United States. Some nations, such as Germany, appear to be tightening their immigration policies, but whether other nations will respond in kind is unknown.[15] Immigration is a major wild card in predicting whether nations will have sufficient younger workers for continued economic growth.

## RUSSIA

Whereas most of the problems of an aging society are the result of better health care and increasing longevity, in Russia the problems are mostly self-induced. Alcoholism is the third-highest cause of death. Mortality rates from heart disease are five to eight times higher than in Western Europe. Russia's population is less than half that of the United States, but traffic accidents account for about 28,000 deaths each year compared to about 32,000 in the United States.[16] About 20 percent of those accidents are attributable to drunk driving.[17] In other areas of manageable risk, Russia ranks alarmingly high. It is the second-largest market for tobacco products after China, even surpassing the United States.[18] The murder count is the ninth-highest in the world.[19] HIV/AIDS is also a problem, with over 1 million cases, more than twice the rate in the United States and Western Europe, and that rate is projected to reach 5 percent of the population by 2025.[20] That would be low compared to Africa, but among the highest of other industrialized nations.[21]

Russia also has one of the lowest fertility rates in the industrialized world, standing at only 1.25 births per woman. Many factors contribute to this, among them poor health care, alcoholism, and high abortion rates. As a result of this lower fertility, the Russian working age population is projected to decline from about 102 million today to 87 million in 2025 and to only 64 million in 2050.[22] Meanwhile, the senior population is poised to explode from 6 million today to 20 million in 2025 and 26 million by 2050.[23] The ratio of workers to seniors is not terribly low, but the health-care issues cause the greatest concern.

If there is a bright side to Russia's future, it would be its abundance of natural resources. It is second only to Saudi Arabia in oil reserves. Other natural resources exist in abundance and are likely to be in high demand in the coming years, such as iron ore, manganese, chromium, nickel, platinum, titanium, copper, tin, lead, tungsten, diamonds, phosphates, gold, and timber—Siberia contains an estimated one-fifth of the world's timber.[24] Russia's ability to export these natural resources will probably ensure that the nation remains a key player in the world economy in the decades ahead.

## DWINDLING POPULATIONS

In the past, futurists and environmentalists warned that increasing populations would eventually doom civilization as we know it. In his 1968 bestseller, *The Population Bomb*, Paul Ehrlich famously warned, "The battle to feed all humanity is over."[25] Ehrlich added that in the 1970s, the world would experience such severe famines, due in part to overpopulation, that hundreds of millions of people would starve. At the time of Ehrlich's warning, global population was 3.5 billion.[26] By 1980, it had reached over 4 billion without catastrophic consequences. Ehrlich wasn't the first to warn about overpopulation. In 1948, when global population was only 2.3 billion, Fairfield Osborn in *Our Plundered Planet* warned of the dangers of continued population growth.[27] We will discuss overpopulation in more detail later, but its dangers have been blown out of proportion. The real problem is not the number of people, but rather human shortcomings. Corruption, incompetence, cultural impediments, inadequate infrastructure, and lack of free-market forces combine to create an inability for nations to support their citizens. For example, population densities are much higher in Europe than in most of the nations of Africa, yet the standard of living is relatively high. African nations have some of the lowest population densities, but hunger and starvation are rampant. Germany has a population density of 229 persons per square mile compared to Ethiopia with only 70, and to the Central African Republic with only 7 persons per square mile. If population density was a primary determinant in well-being and sustainability, one would anticipate that the quality of life would be much worse in Germany than these African nations, but the reality is just the opposite. Nonetheless, aging populations will pose formidable challenges in the coming years.

# PART TWO

# Understanding Aging

With age, virtually every aspect of our physical bodies begins to decline. All five senses begin to degrade to some degree. The speed at which signals are sent from neuron to neuron within the brain and then from the brain to the muscles slows down—we think and react slower. Muscle cells begin to die or lose the ability to function optimally. For a while, exercise can offset this loss of muscle tissue, but age inevitably takes its toll on muscle and bone. That's why you don't see many 60-year-olds playing professional sports.

Curiously, while aging is inevitable, age-related changes can occur quite differently for each individual. Some men lose their hair and others don't. Some see their hair turn gray while others keep the lustrous hair color of their youth. Some people appear to age more rapidly than others while a lucky few seem to be frozen in their 30s, but eventually aging attacks even those lucky few.

Aging is a complex process—so complex that scientists have argued for over two thousand years about what causes it. It's like a disease for which there is no cure. It can be managed, slowed somewhat by diet and exercise, but in the end, it is inevitable—or so goes the conventional thinking.

A big part of the skepticism about treating aging as a disease results from a widespread misunderstanding of aging. Why do we age? What takes place at the cellular level to create each of the individual changes we see as we get older? As we learn the answers to those questions in the following chapters, it will be far easier to understand why a small but growing number of scientists believe that aging is a disease, albeit a very complex one, and one that can eventually be cured.

# FOUR

# Aging and Loss of Function

From the time man invented the first primitive spear, *Homo sapiens* has reigned supreme over all creatures on earth. We have conquered space, walked on the moon, built great machines, produced magnificent art, and created computers that can process a lifetime of calculations in the blink of an eye. Yet over the millennia, man has faced another foe far more formidable than any beast of the plain or forest, a foe that never rests, never sleeps, never stops its inexorable quest to quench the fire of life in every living organism. That foe is aging and, inevitably, death.

Aging attacks so slowly we seldom notice its gradual theft of vitality. It comes silently, taking a little here, a little there with the passage of time until eventually the first subtle signs begin to emerge on the edges of one's consciousness. A ball tossed by a grandchild, so easily caught one-handed years ago, now drops untouched to the ground. A flight of stairs that one flew up two at a time as a child, laughing in pursuit of the next adventure, now requires a pause before reaching the top. We all have seen this, perhaps in our parents or older relatives. For those of us who have already experienced our first gray hairs, we probably have also seen it in our friends or spouses.

Collectively, we might reflect on those younger days and wonder what happened to that boundless energy of youth. These bittersweet memories, fleeting through the minds of older readers even now, are like the taste of a favorite dessert we can no longer eat because of an expanding waistline, or rising blood sugar, or some other age-related malady. Even so, were these the only the pains and troubles of aging, it wouldn't be so bad, but as years become decades, the progression of aging becomes more pronounced.

Eventually, the firm hand that once guided a child years ago now shakes as it holds on to that same child's hand, fully grown to adulthood. The body, once tall and strong, moving through the world in purposeful strides, now moves in halting steps, bent by the withering of muscle and bone. In one of the great ironies of life, roles have reversed. The caregiver has become the receiver of care. Life has come full circle.

In the movie *Indiana Jones and the Kingdom of the Crystal Skull*, Indiana sits with an old friend and reminisces about the recent death of his father and another colleague. The white-haired college professor pauses, sharing Indiana's loss, then quietly sighs, "It seems that we have reached the age where life stops giving us things and starts taking them away."

In truth, this decline starts decades before the twilight years and progresses at different rates for each of us. For some, it becomes noticeable as soon as the 30s—yet others seem to defy the ravages of time until well past what we consider middle age. Even for the genetically gifted, aging eventually progresses to the point where it can no longer be denied.

What causes this inevitable decline? Though the obvious answer is aging, the question then becomes, what causes aging? The answer isn't readily apparent, because aging doesn't appear to have one single cause. It is, instead, the end result of a combination of complex interactions within the body as well as a multitude of external forces, such as industrial pollutants or unhealthy lifestyle choices like poor nutrition or lack of sleep. All of these and dozens, perhaps hundreds, more variables are constantly affecting our bodies every day. To a large extent, how well our bodies adapt to these external and internal stresses determines how fast we age.

## HOW WE AGE

One curious aspect of aging is that it doesn't progress at the same rate for everyone. Most readers have probably endured the humbling experience of attending their twentieth high school reunion only to discover that everyone has gotten older. Not surprisingly, many attendees appear to have aged about 20 years since they were last seen (a refreshing comfort to those of us who have begun to notice the first not-so-subtle signs of aging). Then there is that unlucky person who has aged so rapidly that you wonder if he might be a classmate's father who absentmindedly stumbled into the wrong room, missing his actual classmates by a couple of decades. Yet there are always a few who, from outward appearances, seem to have barely aged at all.

Why do such apparently striking disparities exist among individuals as they age? Obviously, lifestyle, stress, and genetics play a role, but the bigger question is why do our bodies age at all? Why do older adults become more feeble and frail? Why don't we just live to advanced age and, when our functional life span is over, simply fall over dead?

Scientists now know that aging is a lifelong process. Outward manifestations may appear in late childhood but are rarely visible in the first few years of life. If you walked into a kindergarten class of five-year-olds, the children would all appear to be about the same age. Some might be a little taller or shorter than their classmates, but none would appear to be two years old and none would appear to be 15. None would have beards and none would have male pattern baldness—luckily one of those unfortunate aging characteristics that has the good taste not to afflict us until late middle age. It's not until the onset of puberty that we begin to see the first signs of disparities in aging, but even then, the differences are relatively insignificant. Most children begin to reach puberty around age ten and reach full sexual maturity around age 15, give or take a couple of years.

Interestingly, across almost all species, there is a correlation between age of sexual maturity and life expectancy. This even affects subgroups within species. Some ethnic groups and nationalities reach puberty and sexual maturity earlier than other groups. They also tend to have shorter life expectancies, even when corrected for other factors like disease and living conditions.

The predictive nature of life span based on sexual maturity holds true for almost all species and appears to be a pattern that nature has followed for millions of years. The Mediterranean fruit fly reaches sexual maturity in 10–15 days and has a life expectancy of only two months, while the giant Galapagos Island tortoise takes 20 years to reach sexual maturity and has been observed to live to age 177. The lizard-like tuatara of New Zealand—a living fossil whose direct descendants walked the earth with dinosaurs 100 million years ago—is believed to have a life expectancy of 200 years and reaches sexual maturity at age 20. Males have successfully bred in captivity at age 113, and females remain fertile until at least age 80. It seems that when it comes to longevity, sex is a good thing.

Technically, puberty and sexual maturity aren't signs of aging, but they are part of the aging process. As we move through the various stages of life, development and aging follow a natural progression. In the formative years, the emphasis of aging is first on physical development and then on reproduction. The body focuses its growth on the areas it needs to perform optimally at that stage of life.

For most creatures, the end of the reproduction stage marks the beginning of the end of life, but this is where humans differ markedly from other species. It seems that humans continue to grow intellectually far past the reproductive stage of life. Years ago, I noticed that college professors and scientists, some in their 80s or even older, continued to function at an extremely high intellectual level in spite of their advanced age. Rather than losing mental capacity, they were even smarter and wiser than they had been three or four decades earlier. It occurred to me that all humans probably possessed similar potential to varying degrees—they just weren't in professions that rewarded these particular skills or that allowed them to continue working to advanced age.

What would the world be like, I wondered, if all seniors could remain this productive? Granted, there would be some physical limitations, but they shouldn't be insurmountable. A few extra decades of workplace productivity could alleviate some of the most pressing economic problems in the world today, such as the financial crisis of pension and health-care programs for seniors. If seniors didn't retire until their 70s, the funding problems of Social Security would disappear. If they could also stay healthy longer, the burden of senior health care would likewise be reduced. My awakening to the untapped potential for older adults to remain productive into advanced age is one of the insights that led me to write this book. But to change the status quo bias, we must first understand more about aging, including how we see the world.

## HOW WE SEE THE WORLD—OUR FIVE SENSES

As we age we sense the world differently, because most of our five senses decline throughout our life cycles. A typical preteenager can hear sound frequencies as high as 20,000 Hz, but by the time most people reach age 60, the top of the frequency range has decreased to 10,000 Hz. Some clever students have used dog whistle apps to signal others in class, while their professors can't hear a thing! With advanced age, the ability to detect higher frequency sounds continues to decline. One in three 60-year-olds have some hearing loss, while nearly half of seniors age 75 and older have some hearing impairment.[1] Many seniors experience an inability to distinguish words in noisy environments, such as in a busy restaurant.

Over time the sense of taste is similarly impaired, although the change is usually not noticeable until age 60. Once the number of taste buds on the tongue declines from 10,000 to about 5,000, this decreased sense of taste makes food taste bland. This decline varies by individual, but usually salty and sweet tastes are lost first.[2] When nothing tastes good, it's not unusual for the elderly to lose their appetites, resulting in a lack of desire to prepare meals that in turn could lead to malnutrition.

The sense of touch also diminishes with age. As children, our sense of touch is heightened, primarily to discourage us from touching things we shouldn't, like a hot stove. With age, we become more aware of our surroundings and the dangers they pose, and our nerves become less sensitive to pain. This is a good thing, as all of us who have begun to experience the aches and pains of aging would agree. Loss of feeling in fingers in cold weather, by the way, is not caused by loss of sense of touch. In seniors, it's caused by poor circulation that decreases blood flow through the capillaries close to the surface of the skin. Since less blood gets to the toes and tips of the fingers to keep them warm, the fingers get numb. Fortunately, this is one symptom of aging for which many seniors in northern climes have found a cure. They move to Florida!

The sense of smell runs counter to the trend of our other senses. A study of 1,000 Australian males and females found that olfactory functions diminished very little over time.[3] Interestingly, the study found that women had a more sensitive sense of smell than men. Although this study made no attempt to determine why, the difference has not been lost to the perfume industry, which generates about $7 billion in revenue each year.

Among all the senses, vision is arguably the most important in the workplace and, sadly, the most likely to decline with age. Normal eyesight is defined as 20/20 vision—the ability to see at 20 feet what a person with average vision can see at 20 feet. With age, this distance gradually increases. It's not uncommon for bespectacled middle-aged adults to have 20/200 vision, meaning they must be 20 feet away to see what the person with 20/20 vision can see at 200 feet. At 20/400 vision, humans are legally blind.

After the age of 40, we begin losing the ability to focus on items up close due to hardening of the lens inside the eye. Presbyopia, or farsightedness as it is also called, explains why you see people in restaurants holding their menus at arms' length as they try to select their meal. Eventually, even this approach fails and aging adults turn to reading glasses or that dreaded milestone in aging—bifocals. With advanced age, various age-related eye diseases, such as cataracts and macular degeneration, become more prevalent. Also, adults lose one to three degrees of peripheral vision per decade, which is why seniors are more prone to automobile accidents.

By age 70, they have lost about 30 degrees of vision, so they can no longer see vehicles approaching out of the corners of their eyes at intersections.

Of the five human senses, vision is the most potentially limiting to productivity with age, but declining vision is no longer a significant obstacle, in part due to the rapid pace of technological and medical advances in ophthalmology over the past two decades. Except for extreme cases and ophthalmic diseases, most loss of visual acuity can be corrected by glasses or various types of ophthalmologic surgery. In China, for example, over 800,000 Lasik surgeries are performed each year.[4]

## HOW WE MOVE—CONNECTIVE TISSUE

Although the functional limitations of vision and hearing have been mostly eradicated, at least until advanced age or the onset of premature aging due to disease, limitations on physical mobility pose a greater challenge. How we move through the world requires a complex interaction of multiple systems, but the actual movement itself is a function of connective tissue (tendons and ligaments), muscles, and bone. Connective tissue might seem relatively unimportant compared to other major biological systems, but it plays a much larger role. In actuality most of the human body is either made up of it or surrounded by it. Even human bones start out in the fetus as cartilage, which is a type of connective tissue.

Connective tissue is composed mainly of two proteins: collagen and elastin. Our bodies use these two proteins as fibers and fabrics to create a remarkably wide range of tissues within the body. For example, bone and skin—among the hardest and softest tissue in the body respectively—are both composed primarily of collagen, at least in the initial stages of development. It's how collagen bonds with other substances that determines whether it eventually forms skin or bone. In bone, collagen becomes as hard as concrete, while in skin it's stretchy like an elastic band. Our bodies continue to produce collagen throughout our lives, but the production of elastin, a stretchy connective substance that gives skin its elasticity, starts declining after we reach age 15. Perhaps this is why the most visible adverse effects of aging appear first on human skin. Beginning as early as our late teens, microscopic wrinkles begin to form on the face and typically become visible in our 30s or 40s. Obviously wrinkles are caused by aging skin, but why do wrinkles form around our eyes, foreheads, and cheeks? Why do wrinkles form lines at all? Why doesn't skin just shrivel up like a prune or an old peach?

The skin is the largest organ in the human body and is composed of three layers—the epidermis, the dermis, and the hypodermis. Each of these three layers is functionally different, but they are all composed primarily of living cells in a mesh of collagen. Ligaments and tendons are also made mostly of collagen fibers aligned parallel to each other like thousands of tiny bungee cords, but in skin, collagen performs a much different function. Unlike tendons that only stretch along one dimension, skin has the ability to stretch in two dimensions, like a wide elastic band. It's no longer as hard as bone or as strong as tendons, but this structural arrangement of collagen allows the skin to stretch, preventing it from tearing. Even so, collagen by itself wouldn't be sufficiently stretchy to protect the skin from tearing without elastin. Elastin is extremely stretchy, but pure elastin wouldn't be strong enough by itself to make skin. Together, they make skin both tough and stretchy.

Unfortunately, the body drops elastin production by age 15, or sometimes as early as age 13. Without continued production, the body's supply of elastin gradually declines with age. When you pinch the skin on the back of the hand of a child, it will immediately spring back to its original position because the skin has a full supply of elastin. When you pinch the skin on the back of the hand of a 70-year-old, it returns to its original position very slowly. Although other factors, such as inadequate hydration, can affect the speed of this pinch response, the primary variable is loss of elastin.

Dwindling levels of collagen and elastin in the dermis result in more rapid thinning than in the epidermis. This thinning is one of the major causes of wrinkles. As the muscles of the face move beneath the skin, they form creases in the dermis—like a piece of cardboard that has been bent back and forth. The epidermis slides into these creases and manifests an outward appearance of crow's-feet around the eyes, furrows in the brow, and lines around the cheek and jowls. While the first appearance of wrinkles might be devastating to the young thirtysomething adult, they have no bearing on workplace productivity. Indeed, a few wrinkles and a gray hair or two could bestow an aura of experience and wisdom on an otherwise nondescript countenance.

However, we're not so lucky when it comes to loss of functionality due to internal changes in connective tissue. With age, ligaments and tendons lose elasticity, making them more fragile. Unlike skin, which is damaged primarily by loss of collagen and elastin, this compromised elasticity is caused by the binding of collagen with other substances, such as sugar or calcium salt molecules, resulting in a progressive reduction of strength and elasticity. This binding process, called cross-linking, is a major age-related problem in virtually all body systems. In joints, cross-linking causes the normally pliable connective tissue to become thinner and calcified, losing strength and elasticity. The resulting lack of cushioning causes the adjoining bones to develop jagged, irregular calcium deposits in the joint, which is commonly called osteoarthritis.

## MUSCLES

Starting about age 45, we begin to lose approximately 1 percent of lean muscle mass per year as our muscle cells decrease in size and number. While that may not sound like much, cumulatively this age-related muscle atrophy can be significant. It's called sarcopenia and is believed to be the result of damage to the mitochondria—one of the structures within the cell that converts food into energy.

Regardless of the cause, the workforce impact of age-related loss of muscle mass would be limited to jobs requiring manual labor or staying on one's feet all day. However, sarcopenia increases in Western society where large percentages of the population tend to live extremely sedentary lives, resulting in much earlier muscle loss than in more active individuals. The good news is age-related declines in muscle mass can be offset by exercise, which increases the size of the remaining muscle cells. Somehow, for reasons that experts don't fully understand, exercise causes remaining healthy muscle cells to compensate for the atrophied fibers and lost cells.

While this is great for adults who actually exercise, a U.S. Health and Human Services study found that only 20 percent of adults exercised regularly.[5] Worse, only 13 percent of seniors aged 65–74 exercised, and exercisers dropped to only 6 percent by age 75. This is unfortunate because age-related loss of muscle functionality can be reversed through proper exercise. A Tufts University study found, for example, that after a 12-week exercise program, older adults aged 56 to 81 added 3 pounds of muscle, lost 4 pounds of fat, and significantly improved their strength.[6]

However, even exercise cannot forestall the erosion of aging indefinitely. Over time, the remaining cells begin to lose function regardless of continued exercise. Studies of 80-year-old bodybuilders found that even those who continued to exercise with the same intensity gradually lost muscle mass with age, despite maintaining exceptional muscle definition. Fortunately, the stereotype of weak and frail senior citizens does not have to be a reality. The functional decline of muscle tissue can be delayed until an advance age with something as simple as a basic exercise program. Ernestine Shepherd, over the age of 75 and the oldest competitive female bodybuilder in the world as declared by the *Guinness Book of World Records*, didn't begin training until age 71. Just one look at online photos of "Miss Ernie," as she is known in professional bodybuilding, is worth a thousand words.

### BONES

Although bones begin as cartilage in the fetus, by birth collagen begins binding with calcium and other minerals to form 28 subtypes of collagen proteins, some of which form bone. It's hard to believe the ear of a newborn child, so translucent you can actually see through it, begins as the same cartilage that forms the infant's bones. Even the newborn baby's soft and smooth skin begins in the same way. Yet over time, bone will become so strong that it will rival concrete in strength at only a fraction of its weight. In fact, bones are so hard that Stone Age man carved them into primitive weapons and tools.

The growth of bone in the human body is kept in balance by two opposing cell groups: osteoblasts and osteoclasts. Osteoblasts produce collagen, which in turn binds with calcium and other minerals to form bone. Osteoclasts, on the other hand, dissolve bone. Throughout infancy and childhood, osteoblast activity is greater than the destruction created by osteoclasts, but around age 40, the level of activity reverses. The digestion of bone by osteoclasts exceeds the growth of new bone by osteoblasts, which results in a gradual bone loss with age.

This decline is called osteoporosis and is considered both a disease and a natural age-related activity, depending on its onset and severity. Although theories exist, scientists aren't really sure what causes osteoclasts to become more active than osteoblasts with age. But it appears that almost all age-related bone loss results from the imbalance of these two opposing forces. If scientists could find a way to maintain a balance between osteoblast and osteoclast activity, it would, in theory, stop age-related bone loss forever.

## THE BRAIN

The enormous capabilities of the human brain make our species unique among the creatures of this planet. In the absence of age-related diseases, we now know that bones and muscles can retain most of their functionality until advanced age. If the brain could stay healthy as well, seniors could reasonably expect to remain productive into their 80s.

One of the greatest misconceptions of aging is the belief that we face a significant loss of mental capacity with the passage of time. Older adults jokingly refer to forgetfulness as having "a senior moment," but such momentary lapses of recall are only one characteristic used to gauge mental capacity. Scientists measure cognition in several distinct categories, from areas as simple as attention span and memory to higher-level cognitive functions, such as language and complex decision making. Within these major categories there are several subcategories. Seniors have proven they can perform as well as or even better than younger adults in certain categories. For example, seniors may have a broader base of knowledge to draw from (semantic memory), or they might be less prone to risk-taking or emotional decisions.[7]

Unfortunately, there are other areas of mental function that decrease with age. For instance, the ability to multitask declines, and the pace at which seniors learn new information slows. Researchers once believed that seniors had a diminished capacity to acquire new knowledge because they fared poorly on tests when given a time limit to learn new material. However, when subjects were allowed to learn at their own pace, the disparity between seniors and younger adults disappeared.

Until recently, scientists believed that the human brain reached full maturity by age 40 and then began an inexorable decline due to loss of brain cells. They further believed that this loss of brain cells led to memory loss and eventually to dementia or Alzheimer's disease in some seniors. Today, it has become clear that, in the absence of such age-related diseases, brain cells can remain healthy and functioning throughout our lives. In fact, scientists now know that the brain continues to produce new nerve cells, called neurons, well past age 60. In theory, this would imply that the brain could continue functioning at very high levels indefinitely since new brain cells are continually produced, but the reality is that most seniors eventually begin to lose some cognitive functioning. There is no question that with age, brain reaction times slow.

In a healthy young brain, when neurons send an electrical impulse to the muscles to react, the signals travel along link-sausage-like structures called axons to the muscles. Axons are surrounded by layers of a fatty substance called myelin, and these layers are called the myelin sheath. Together, axons and their myelin sheaths look like a coaxial TV cable and accelerate the electrical signal so it reaches the muscles much faster. With age, the thickness of the myelin sheath decreases, diminishing the speed at which signals are sent to and from the brain. Some believe that decreased myelin lies at the root of all age-related declines in the normal brain, while other scientists believe it is only one of several causes. Regardless, it is certainly a major contributor to age-related decline of mental capacity.

It's quite possible that most of what is perceived as a *normal* age-related decline in mental function isn't normal at all. It may well be the result of any number of diseases that impair brain function, such as senile dementia, diabetes, chronic alcohol consumption, congestive heart disease, blocked arteries, and of course Alzheimer's disease. After age 65, the prevalence of Alzheimer's doubles every five years—by age 85 Alzheimer's disease affects roughly half the senior population.[8] What causes Alzheimer's is unknown, but there are two primary suspects: amyloid plaques and neurofibrillary tangles.

Amyloid plaque is a protein-based substance that accumulates between the neurons in the fluid matter. Since it lies outside the neuron cells, amyloid plaque is called extracellular matter. Until recently, it was widely believed that amyloid plaques caused Alzheimer's disease by interfering with the transportation of nutrients and waste materials to and from the cells. However, significant buildups have been discovered in the brains of otherwise healthy seniors, which led other scientists to speculate that these plaques might be a protective reaction, perhaps serving to isolate a more poisonous soluble form of amyloid.

Neurofibrillary tangles, on the other hand, build up inside the brain cells and are thus a type of intracellular matter. These tangles are caused by the breakup of microtubules, the tubelike transport system that moves nutrients from one area of the cell to another. For some unknown reason, these microtubules collapse with age. Then tau, the type of protein that makes up the microtubules, twists into insoluble fibers that cannot be transported out of the cell. These neurofibrillary tangles clog the neuron, prevent normal operation, and eventually result in cell death.

As Alzheimer's disease progresses, brain cells die and the brain shrinks, neurofibrillary tangles and amyloid plaques increase, and brain functions decline precipitously. Eventually Alzheimer's patients are unable to recall their own names or the names of their children or spouses. Regardless of the cause, the disease can be devastating.

In the absence of these diseases, it's an entirely different story. Many categories of cognitive function show no decline at all or at least no decline until extremely advanced age—speech, language, certain types of memory, nonemotional decision making.

Among the other categories of mental function that do decline—ability to multitask, for example—there is a wide disparity from individual to individual. Seniors have also demonstrated an ability to compensate for declining mental function in one category by using their strengths in another area. For example, an elderly college professor might forget a word, but because his vocabulary is so large, he can quickly replace it with another word without anyone being the wiser. The one cognitive function that always declines with age is speed of reaction, but many seniors develop compensatory skills to hide this decline, such as anticipating questions in advance or using filler phrases to buy time to think of the proper response: "Well, let's see . . . (pausing). I think we should . . ."

In a study spanning over 50 years, Mills College psychologist Ravenna Helson, now a professor at UC Berkeley, tracked the mental and emotional functions of 123 adults. Surprisingly, she found that the brain actually continues to grow in its ability to process certain types of information. The complex reasoning and mental/emotional balancing skills of many didn't peak until their 50s or 60s. Relationship skills also improved.[9]

A seven-year longitudinal study of cognitive decline with age found that only perceptual speed began to decline in young adulthood. Most brain functions continue to grow until the 60s and are still around the same levels at age 70 as at age 25. Verbal ability and inductive reasoning are about the same at age 88 as at age 25.[10]

In younger adults the right and left hemispheres of the brain operate somewhat independently—the left brain handles logical functions much like a computer and the right brain handles intuition, creativity,

# Decline in Cognitive Functions with Age

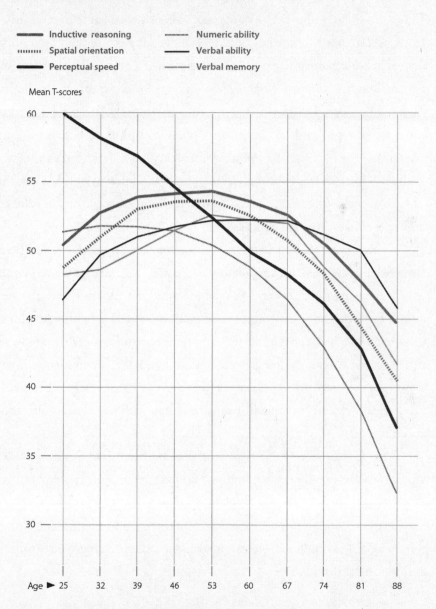

Based on seven-year longitudinal data from from the Seattle Longitudinal Study.

K. Warner Schaie, et al, The Seattle Longitudinal Study: Relationship Between Personality and Cognition, Neuropsychol Dev Cogn B Aging Neuropsychol Cogn. June 2004; 11(2-3): 304–324.

*Figure 4.1. Decline in cognitive functions with age.*

and other nonlinear types of thinking. Curiously, as we age, the communication channels between these two hemispheres of the brain improve. In tests of older adults performing tasks requiring multiple mental functions, researchers found that, unlike younger subjects, older adults often activated both hemispheres at the same time when evaluating information and making a decision.[11] Roger W. Sperry, an American psychobiologist, received the Nobel Prize in 1981 for his discoveries relating to the functions of the right and left brain, but poets and philosophers have long recognized the older brain's innate ability to balance logic and experience with emotions and perspective. "The young man knows the rules, but the old man knows the exceptions," quipped nineteenth-century physician and poet Oliver Wendell Holmes.[12]

It's even possible that the extent of mental decline with age has been somewhat overstated. It has not been practical to test the same people over a time span of 50 years, so studies of mental function actually test adults from different generations. Researcher Timothy Salthouse points out that what was considered to be age-related differences in cognitive function could actually be a generational difference.[13] When mental functioning *of the same people* was measured over the years, they either stabilized or showed an increase in brain function with age at least up until about age 60.

What is clear from these studies is that, barring mental disease, older workers can be productively employed far past the current retirement age. Skeptics have only to browse freelance websites like Elance.com or Freelancer.com to find thousands of workers past their retirement age bidding for reasonably complex projects requiring excellent inductive reasoning and writing skills.

## HOMEOSTASIS

Earlier, we learned how osteoblast and osteoclast activity remained in balance until middle age. This is an example of homeostasis—the body's ability to regulate its inner balance in response to outside stimuli. Another example is the proper regulation of body temperature, which is critical to normal function. When the outside temperature is hot or when we exercise and create excess body heat, the body maintains its internal temperature balance by sweating. Conversely in cold weather, it attempts to stimulate a higher body temperature by shivering or raising goose bumps, the latter of which erects hairs that trap air, creating a layer of insulation.

One of the most challenging problems of aging is the decreasing ability of the body to maintain homeostasis. Each system in the body operates within a narrow range of acceptable temperature and chemical conditions. If an external stimulus causes the body to move outside that range and homeostasis cannot be reestablished, the results can be quite severe and include disease, permanent damage to organs, or even death. Homeostasis becomes increasingly important for seniors because one system tries to compensate for the diminished capacity of another system, and that in turn creates problems. For example, partially blocked arteries cause the heart to pump harder to get blood throughout the body. The harder pumping increases blood pressure, which increases the risk of stroke.

### THE CARDIOVASCULAR SYSTEM

The cardiovascular system carries oxygen and vital nutrients to cells throughout the body and then carries waste products from these cells to the lungs, liver, and kidneys, where they are eliminated. During exercise, the heart speeds up to maintain homeostasis, pumping oxygen and nutrients to the muscles faster.

With age, this vital transportation system becomes clogged like a freeway after an accident. Collagen cross-links with calcium and sugar, which leads to the narrowing of the tiny capillaries that carry blood and nutrients throughout the body. This cross-linking of collagen, sugar, and calcium as well as cholesterol deposits creates plaque buildup in veins and arteries, making the walls less pliable, further restricting the flow of blood. In order to maintain homeostasis and send the same amount of oxygen and nutrients to the cells throughout the body, the heart compensates by pumping harder, which can raise blood pressure to dangerously high levels. The plaque buildup also decreases the diameter of the lumen of the blood vessels (the inside space of the vessel), contributing to the increase in blood pressure.

Furthermore, collagen cross-linking makes the larger arteries stiff and hard, which also raises blood pressure. Elastic arteries act as shock absorbers for the heartbeats' pressure pulses, lowering the pressure as they stretch and rebound. A stiff arterial wall transmits the high pulse pressure unabated to the brain, where the incessant pounding can burst a blood vessel, resulting in stroke.

The aging process also affects the heart itself. As the heart walls become thicker, they lose the ability to pump as much blood volume with each beat. As the arteries to the heart itself become blocked, the lack of blood flow to the heart can cause angina pain or a heart attack. If blood flow to the brain becomes too restricted, it can result in a stroke, the third leading cause of death among seniors.[14]

None of the body's systems are spared the effects of aging. Lung capacity decreases as the intercostal muscles and chest wall harden with age. The pancreas loses its ability to regulate blood glucose levels, often resulting in adult-onset diabetes. The digestive system loses its ability to absorb nutrients efficiently, causing indigestion or acid reflux. Hormone levels can drop precipitously, not just in postmenopausal women but also in men. It is not uncommon to see an otherwise healthy 70-year-old man with testosterone levels 50 percent lower than that of a young adult male. Among men who are sedentary or overweight, the drop can be even more severe.

Earlier we learned that as long as we stay healthy, the loss of function in bones, muscles, and connective tissues is not a major limiting factor in workplace productivity. Similarly, in the absence of Alzheimer's disease and senile dementia, a healthy brain can function productively to advanced age. So, not only is it possible to remain in the workforce until advanced age, it's highly desirable for everyone involved. Employers benefit by retaining the wisdom and experience of older employees. Nations could benefit by delaying eligibility for pension plans. And since pension plans would be postponed, individuals would benefit by being able to remain gainfully and productively employed until they reached the new retirement age or saved enough to retire.

Nonetheless, it's clear that increasing age brings on a decline in the optimal functioning of our body systems. Whether the cause is natural aging or premature aging brought on by disease, decreasing functionality with age creates major challenges for employers and governments. Slowing the aging process—or better yet, stopping it entirely—would provide immeasurable benefits to society, but first we need a greater understanding of what causes aging at the cellular level.

# FIVE

# Biologic Aging
# at a Glance

"**W**hy is Grandpa so old?"

It's a question that every three-year-old child has uttered at one time or another. Parents usually just laugh and glance knowingly at each other. "Grandpa is old because he is, well, old," they respond wisely. "Getting old is just one of those things that happen."

It's a reflexive response given without much thought, but if we pause to seriously consider the question, the fallacy in such circular logic becomes evident. Sure, we age as we get older, but why does getting older mean that we show the telltale signs of deterioration commonly associated with aging? Or, to put it to a finer point, what causes aging?

Since most people love a good murder mystery, let's recast death due to old age as the oldest murder mystery of all time. Whodunits have long been one of the most popular genres of literature. Our older readers can probably remember spending childhood summers reading the adventures of the Hardy Boys or Nancy Drew. Everyone has seen screen adaptations of Sherlock Holmes or Agatha Christie's classic stories. Many of the most popular shows on television have been murder mysteries—*Murder She Wrote, Perry Mason, Columbo, CSI, The Closer, NCIS*, and *Bones*, just to name a few.

Over the years, many suspects have been identified as the primary causes of aging and death. As a result, many "theories of aging" have risen and fallen, discredited by advances in medical science as new knowledge emerged. Still, some scientists remain dogmatically attached to their pet beliefs on aging. It's reminiscent of the retired Belgian detective Monsieur Bouc in *Murder on the Orient Express* who would identify a suspect and then dismissively declare that the search was over . . . he had found the culprit and the case was closed. Of course, each of these theories of aging have circumstantial evidence to support them, but as Hercule Poirot cautioned Monsieur Bouc, "No, it is hardly so simple as that." Meanwhile, the aging research world's equivalents of the world's legendary sleuths—Agatha Christie's Hercule Poirot, Arthur Conan Doyle's Sherlock Holmes, Jessica Fletcher of *Murder She Wrote,* and even the cigar-smoking Lieutenant Columbo in his crumpled raincoat—have been investigating the unexplained deaths of seniors for years. The clues were numerous, the suspects many, and the plot convoluted with classic red herrings. But in the end, the villain—or in this case, villains—have finally been discovered.

Were our modern-day detectives to gather all the suspects into the library and lay out the case, what would they reveal about the causes of aging? Would it be Colonel Mustard in the kitchen with a candlestick? Or could be the sultry Ms. Scarlet in the study with a spanner (a wrench in the U.S. version of Clue)?

## CAUSES OF AGING

The debate over the causes of aging has been going on for quite a long time—so long that, not surprisingly, scientists have identified several plausible "theories of aging," each believed by their respective investigating detectives to be the villain. But as you will soon see, not only were there multiple suspected causes of aging, the causes frequently overlapped with one another, leading scientists to wonder in bewilderment, which was the cause and which was the effect? Was a particular mechanism the primary cause of aging, or was it an effect of some other mechanism, perhaps another yet-to-be-discovered process?

In reality, aging is a complex, multidisciplinary process with a myriad of causes. When a tire fails on a car, it could be from any number of reasons—poor alignment, low air pressure, excessive braking, sharp objects, potholes, or excessive mileage. Not all tires wear out for the same reason. It's usually a combination of factors, with one final factor being the determinate in that particular situation. It's the same with aging. When a person dies after reaching old age, there are usually multiple contributing factors even though one primary factor might appear on the death certificate.

In order to find ways to slow or even reverse the aging process, scientists assumed that the first step would be to identify potential causes of aging. This may likely be an incorrect assumption, as we'll discuss later, but nonetheless it's instructive to learn a bit more about the causes of aging before we discuss some of the exciting ways that medical science is discovering to combat them. Since there are multiple factors in aging, the following are listed in no particular order.

## ENDOCRINE DAMAGE

The endocrine system influences the body at every life stage—puberty, sexual maturity, physical growth to adulthood, and menopause. Its effects are predetermined and driven by changes in hormone levels. After the end of the reproduction stage of life, hormone levels drop. This change occurs on a predetermined schedule around the mid-40s and early 50s in women, resulting in the onset of menopause and an increased risk of age-related diseases, including heart attack and stroke. The decline in hormone levels in men is less sudden but just as dramatic over time. By age 70, testosterone levels have dropped 75 to 90 percent in males.

The clues supporting a hormonal influence on aging are seemingly everywhere. Loss of muscle mass and strength is evident with age. The symptoms of menopause have created a multibillion-dollar market for hormone replacement therapy. The rush for hormone replacement for males is equally apparent with the advertisements addressing "Low T," a euphemism for low testosterone. Further anecdotal evidence of measures to combat declining testosterone levels can be found in the pharmaceutical industry. Studies using hormone supplementation show that returning hormones to youthful levels slows some of the negative impacts of aging but does not slow the overall aging process or extend life. In fact, in some cases artificially increasing hormone levels has been shown to increase the incidence of age-related diseases.[1] Still, hormones certainly affect health and overall fitness, creating the outward appearance of influencing aging, but whether this is a direct result of the hormones themselves or something else remains a mystery.

## GENETICS

Our genes play a very important role in longevity and control how we cope with aging. People with long-lived close relatives are expected to live longer. Families with certain genetic traits may have lower risk of age-related diseases and may even look younger and remain more vigorous in advanced age. Most of the genetic differences associated with increased life expectancy deal either with the rates of metabolism, activity, and the stability of the repair mechanisms and robustness of the bones, organs, and tissues. Genetics control the aging of animals in a similar way as the design controls the aging of cars. Some cars are built to last longer with robust components, combustion control, diagnostics, and repair features. We can make cars last for decades, but we cannot switch on a program to stop aging. We need external maintenance, repair, and replacement services to make it run for a very long time. The difference between animals and automobiles is that animals are designed to mature and produce offspring. After that they are no longer needed. Some animals—certain species of rodents and female salmon, for example—die almost immediately after producing offspring, but most get a grace period to take care of the young while their regeneration and maintenance mechanisms degrade and fail.

Interestingly, for most species there is a direct correlation between the time it takes to reach sexual maturity and life span. Humans typically experience the onset of sexual maturity around age 13 and have a maximum life span of 122. Most other species show a similar correlation between the time to reach sexual maturity and overall lifespan. Creatures that reach sexual maturity in months typically are very short-lived, whereas species that take many years have much longer life spans. Many other mammalian species have ratios of age of sexual maturity to life span that are similar to that of humans—mice, monkeys, and elephants, for example.

Correlation, however, does not imply causation. As Holmes admonished Watson, "Never theorize before you have data. Invariably, you end up twisting facts to suit theories, instead of theories to suit facts." In 1984 Michael Rose, an evolutionary biologist at the University of California, Irvine, proved that the ratio between sexual maturity and life span was not fixed. Rose selectively bred fruit flies at their earliest age of reproduction and others at their latest age of reproduction. He found that when reproduction was delayed to late stages of sexual maturity, lifespan increased dramatically in future generations without a corresponding increase in time to reach sexual maturity in subsequent generations.[2] If genetics were the sole cause of aging, then the age to sexual maturity would have also increased. Also, no one has been able to demonstrate how the mechanisms controlling genetics could influence age-related changes.

## THE HAYFLICK LIMIT

In 1961 Leonard Hayflick discovered that human cells in a Petri dish would only divide 40 to 60 times before cell division stopped.[3] This was a major discovery, because at that time, the widely held belief was that human cells could keep reproducing themselves indefinitely. The "Hayflick Limit" appeared to prove that each human cell, or to be more precise, the DNA within those cells, contains a biological clock permitting them to divide only a predetermined number of times. Since the Hayflick Limit can only be observed outside the body—such as in a Petri dish—there is ample support to argue that the finite ability of the cells to divide is not predetermined, but instead is an artifact of no longer being within the body. Even if the Hayflick Limit does apply to cells within the body, it still leaves scientific sleuths with another riddle to solve. What caused the cells to stop dividing? This unsolved mystery eventually led anti-aging detectives to investigate the causes of DNA damage and its effect on aging.

## DNA DAMAGE

Each strand of DNA contains thousands of genetic markers—the codes of life that make each of the 7 billion people on earth uniquely different from one another. Everything from the color of our hair and eyes to whether our middle toe is longer than our big toe is determined by the genes held in our DNA. Healthy DNA is essential for cells to divide accurately, keeping the codes of life intact within the DNA from one replication to another.

The iconic double helix of DNA is familiar to many people. It looks a bit like two ropes wrapped around each other to form a longer, thicker rope. At each end of every strand of DNA is a protective structure called a telomere, composed of DNA and special proteins. This telomere structure has a very specific sequence of DNA code, which prevents the sticky ends of chromosomes from adhering to each other and interfering with cell division.

Over time, the telomeres gradually shorten, eventually exposing the sticky ends of the DNA. As a result, when the DNA attempts to replicate itself, the ends stick together, which results in chromosome breaks and incomplete replication. The body's natural repair enzymes attempt to fix these damaged DNA strands, but for reasons that aren't fully understood, the body's ability to efficiently repair this type of damage declines with age. Eventually the rate at which damage occurs can overwhelm the body's capacity to repair it. When this happens, the cells stop dividing, die, or in some cases, mutate and then replicate very quickly, leading to cancer.

While DNA damage increases with age and certainly appears to contribute to aging, it remains unknown whether it is a primary cause of aging or the effect of some other mechanism. In addition, scientists have yet to find a clear correlation between aging and the amount of DNA damage found in tissue. Once again, this villain must have other partners in crime.

## METABOLIC RATE

In 1908 physiologist Max Rubner discovered a relationship between metabolic rate and longevity. Smaller animals have shorter life spans and faster metabolisms than larger animals. In other words, larger species that expend fewer calories per gram of body mass tend to be more long-lived than smaller, short-lived species. Rubner surmised that all animals were born with a finite amount of potential energy; the faster they used it, the faster they died.[4]

Rubner's observations appeared to be true across the board in 1908, but as time passed, researchers uncovered some notable exceptions. The hummingbird, for example, has a metabolism that is 100 times faster than that of an elephant. Since the maximum lifespan of an elephant is about 55 years, it would seem that a hummingbird life span should be a little over six months, yet hummingbirds have been known to live for over 12 years.[5]

More importantly, Rubner's premise is less conclusive in humans. According to his discoveries, sedentary persons should live longer than those who exercise regularly, yet numerous medical studies suggest the opposite is true. Of course, there are other mitigating factors. The rate of obesity increases dramatically in sedentary populations, resulting in increased risk for disease and premature aging.

## FREE RADICALS

Free radicals and oxidants—jointly referred to as reactive oxygen species (ROS)—are atoms with an unpaired electron. Electrons prefer to orbit an atom in pairs, so they scavenge another electron from a productive cell. The productive cell can no longer fulfill its purpose. Over time, enough free radicals can cause damage that accumulates with age. Although free radicals are created by any type of stress, as well as by environmental pollutants, they are primarily a natural byproduct of mitochondrial energy production within each cell. Normally, these byproducts are eliminated by the body's natural antioxidant defenses, but they can overwhelm these mechanisms during times of physical, emotional, or environmental stress. These free radicals often have an electric charge, making them like little magnets. As proteins and molecules pass by in the process of their normal duties, free radicals attach themselves to connectors on their surface, preventing them from carrying nutrients to the cells or carrying garbage away from the cells. In some cases, free radicals cross-link two larger molecules together, making them too large to be transported through the cellular membrane, or otherwise interfering with normal cellular function. This harmful process is a major form of age-related damage in the body, causing hardening of the arteries, plaque buildup in arterial walls, and damage to heart muscle.

There's little doubt that increased free radicals have a negative impact on health, increasing the risk of many age-related diseases, but the link to aging itself is less clear. As Holmes observed to Watson, "Circumstantial evidence is a very tricky thing. It may seem to point to one thing, but if you shift your point of view, you may find it pointing to something entirely different." Indeed, attempts to alter lifespan by alternately reducing or increasing free radicals have met with mixed results. Experiments with fruit flies tend to support a connection between free radicals and aging, while experiments in mice and other vertebrates yield mixed results. There's also a question of cause and effect—are free radicals the *cause* of aging, or are they a symptom of some other mechanisms in play, such as hormones or metabolism, each of which has the capacity to increase free radicals?

## OTHER AGING THEORIES

There are even more theories about what might be major contributing factors in aging. For instance, waste products—specifically lipofusin, a gunky mix of lipids, sugars, and even metals—build up inside cells' recycling compartments, or lysosomes, preventing the cells from replacing damaged structures. Another possible contributing factor to aging is the gradual loss of the thymus gland, which may, in a sense, act as a biological clock. Another contributing factor in aging is the cumulative damage caused by the binding of sugar and collagen proteins, called glycation, which causes hardening of the arteries and heart muscle, among several other types of damage. When seniors survive past age 100, glycation is very often a primary factor in their eventual deaths, but it appears to be less of a factor in earlier deaths, except among diabetics. The obvious question is what causes the deaths of the 99 out of 100 seniors who die before age 100? In short, there are many possible causes of aging, but none of the ones we've discussed so far seems to be the overarching factor.

## EVOLUTION

Evolution has also been cited as a major contributing factor in aging. Academic luminaries such as Peter Medarwar, George Williams, Thomas Kirkwood, and W. D. Hamilton have all cited evolutionary processes to explain why mortality rises with age. Michael Rose, an evolutionary biologist at the University of California, Irvine, and author of *Evolutionary Biology of Aging*, sees aging as an unintentional byproduct of the evolutionary process. He suggests that the process of natural selection is the primary driving force in life. Natural selection ensures that genetic mutations optimizing our survival and fertility are passed on through the eons. This process favors gene variants beneficial to survival and reproduction early in life but which might unfortunately produce deleterious effects later in life. There was no need to build defenses against these late-life damage mechanisms because by the time the damage became pathological (causing health problems), after age 30 or so, primitive humans would have most likely fulfilled their primary function—which is to reproduce and pass on desirable traits.

The human body is the equivalent of a finely tuned Formula One race car. Over the years these cars evolved with three goals in mind: go fast, handle superbly, and survive long enough to finish the race, which typically lasts for three hours or less. Imagine for a moment what it would be like if, after the race, these drivers kept using these cars year after year as their primary vehicle—commuting to work, picking up groceries, taking the kids to school. Eventually the same features that were specifically selected over the years to give these race cars a competitive advantage could lead to their downfall. The brakes, having only enough padding to complete a three-hour race, would eventually begin to fade. The radiator, designed to provide cooling at speeds in excess of 200 miles an hour, would suffer stress in stop-and-go rush hour traffic. Eventually, this highly tuned racing machine would succumb to forces it was never designed to handle.

This is similar to what happens with humans. Evolution has optimized humans for early survival, reproduction, and a sufficient life expectancy to raise offspring to maturity. Once those primary roles are accomplished, the body survives as best it can.

## REPAIRING THE DAMAGE

Rather than focus on the causes of aging, perhaps there is a better approach that bypasses causes entirely. In 2001, while attending a conference on aging, biogerontologist Aubrey de Grey—one of the world's leading authorities on aging—had an epiphany. The focus on determining a cause of aging was distracting his colleagues from a far more important issue—how to repair the damage leading to aging and age-related diseases.

In many cases, it's easier to repair damage than it is to figure out exactly why the damage occurred. For example, scientists don't know exactly what causes all the underlying mechanisms that combine to cause farsightedness with age, but now it can easily be corrected with LASIK surgery. Although there are theories, we don't know exactly why arterial plaque builds up in some people but not in others, but we have multiple ways to repair the damage. If it's not necessary to know why damage occurs in order to repair it, de Grey theorized, it's not unnecessary to understand why aging occurs in order to treat it.

In 2007, de Grey wrote a landmark book to bring his revolutionary ideas into widespread awareness. *Ending Aging* explains how all aspects of aging and disease-related death fall into one of seven classes of damage that should eventually be repairable. He developed a comprehensive plan, called "strategies for engineered negligible senescence" (SENS)—that is, strategies to slow or stop the aging process by engineering new medical advances. SENS identifies seven major classes of damage and describes approaches to fix each one.

Aubrey de Grey is a colorful and, some would say, eccentric chap— British, tall, thin, and graced with a flowing red beard that reaches more than halfway to his belt. Dressed in his customary faded jeans and tennis shoes, de Grey stands out at the many international events where he is a featured speaker. Whereas other speakers often list hobbies such as hiking or swimming, Dr. de Grey lists his interests as "studying the accumulated and eventually pathogenic molecular and cellular side effects of metabolism," or to put it in a bit less formal terminology, studying the damage that causes aging.

He further believes that we can develop the technology to repair all seven classes of damage within the life expectancy of today's youngest generation, and helped found the SENS Foundation to catalyze the development of the new class of medicines based on this approach.

A primary goal of the SENS Foundation is to develop techniques to maintain health and improve functioning in older people. These health-improving breakthroughs should build upon each other until we reach a point that de Grey calls "longevity escape velocity"—the point at which damage can be repaired fast enough to keep subjects alive and healthy until the next round of medical discoveries further extends life span. Once longevity escape velocity is achieved, humans would be on the path to achieve a succession of rejuvenation therapies, each powerful enough to extend life expectancy even further, eventually rendering people impervious to the ravages of aging and death except in the case of accidents or violence.

The human body is like a machine, albeit a very complex machine, and as such it can be repaired without fully comprehending why the damage occurred. For example, you might not understand what happens in your car at the molecular level to cause particulates to accumulate in the oil or why these particulates cause damage to your car's engine, but it's not necessary to understand why the damage occurs to prevent it from happening. You (or your mechanic) can simply change the oil before the damage reaches pathologic levels. While the damage-related mechanisms of the body are certainly more complex, the same principles apply.

To extend the machine analogy further, we can look to brakes to see how modern medicine approaches illness. Brake pads are made with a thick layer of abrasive material bonded to a metal subsurface. When the brakes are applied, this abrasive surface is pressed against a metal plate called a rotor, which causes the wheel to rotate slower and eventually stop. With repeated use, this abrasive material wears off, exposing the metal body of the brake pad. When the metal subsurface of the brake pad contacts the metal on the rotor, it gives off a high-pitched squeal. Many people wait until their brakes start making noise to take their car to a repair shop, but by then it is too late to simply replace the brake pads— the rotor must also be replaced at substantial expense, the equivalent of major surgery on the car. Similarly, many people wait until an age-related disease manifests itself with signs that are so obvious they can no longer be ignored without seeing a doctor. As a result, invasive techniques like surgery or chemotherapy are the only options that remain.

The SENS approach is different: allow the damage to occur naturally but intervene before it reaches pathological levels—that is, repair the damage *before* it causes a disease. Returning to the automobile analogy, this would mean periodically checking the brake pads and replacing them before they begin to damage the rotor. Similarly, preventive medicine keeps us healthy for a few years longer so that we can benefit from the next round of medical breakthroughs, giving us a few more years to survive.

## EXTRACELLULAR AGGREGATES

Damaged proteins outside the cell can change shape. These fibers build up in the area surrounding the cells, interfering with the function of mechanisms responsible for transporting nutrients to and waste products away from the cell. Eventually this interference can become so great that it hinders the normal functioning of the cells. Furthermore, many of these amyloids and aggregates become sources of harmful free radicals.

Although the most widely discussed of these amyloids is beta amyloid plaque (discussed earlier as a primary cause of Alzheimer's disease), a different type of amyloid buildup is believed to be a factor in Parkinson's disease. Still another amyloid increases the risk of a form of heart disease known as senile cardiac amyloidosis. With advanced age, amyloid buildup becomes an increasingly common cause of death. By age 90, over half of seniors exhibit senile cardiac amyloidosis, in one study of seniors living beyond the age of 110, amyloidosis was the cause of death in four out of six who died during the study.[6] There are two hypothetical strategies for preventing amyloid damage. One approach would be to create a vaccine or an antibody that either removes amyloids or prevents them from forming in the first place. In theory, a vaccine could work forever, whereas antibodies would have to be periodically injected every few years to boost the body's defense mechanisms against amyloid buildup. The second approach would be to develop mechanisms that attack these bound-together proteins and break them up so they could be swept away by the body's natural waste disposal systems.

## INTRACELLULAR AGGREGATES

As the name implies, intercellular aggregates are materials that build up inside cells. Within the cells, unneeded or damaged structures are sent to lysosomes, sort of a recycling center, to be broken down into raw materials that can be used to create enzymes, repair cellular membranes, and create raw materials for other natural functions. For a variety of reasons, over time lysosomes lose their ability to break down waste products into useful materials. In short-lived cells, this problem would resolve itself as cells die and are replaced by new cells, but in long-lived cells, such as in the brain and the heart, this intracellular junk builds up inside lysosomes until they can no longer function properly. Most of this intracellular junk is lipofuscin—a gunky substance composed of oxidized fatty acids, sugars, and metals. Lipofuscin and other intracellular aggregates are considered to be contributing factors in atherosclerosis, heart failure, Alzheimer's disease, senile dementia, and macular degeneration.

This buildup occurs because lysosomes evolve to handle the most prevalent cellular waste products but *not every possible waste product.* A basic precept of natural selection is that the body makes very efficient use of resources. Disposing of less common waste products that take decades to reach toxic levels was simply not a priority because Stone Age man seldom survived long enough for these accumulated wastes to be a problem.

## CELL LOSS

Some cells die and are replaced every few weeks, such as red blood cells. Others last an entire lifetime, like brain and nerve cells. When these long-lived cells die, the remaining cells attempt to shoulder the workload, but that can't go on forever. Eventually, the cumulative cell loss can result in disease or death.

Genetics can predispose cells to early death, but other environmental factors—DNA damage, free radicals, buildup of intracellular aggregates, and viruses, just to name a few—play a role as well. Cell loss is a primary factor in Parkinson's disease, multiple sclerosis, and Huntington's disease. Since the cells are dead, there's no way to repair them—they must be replaced. Either an individual's own stem cells must be stimulated to divide and create replacement cells, or some other form of stem cell therapy would be required.

## DEATH-RESISTANT CELLS

Within the human immune system, T-cells (T-lymphocytes) are on the front lines to defend us against viral infections. There are two primary kinds of T-cells: naïve T-cells and memory T-cells. The former analyze viral attacks and modify themselves to best defeat viruses, then they replicate themselves and attack. After reproducing large numbers of T-cells to beat back the enemy, the immune system must rebalance itself, as the body carefully manages its resources and only allows the immune system to keep a certain amount of T-cells in reserve. As a result, the body destroys the excess T-cells that are no longer needed to fight the virus.

Not all are destroyed however. A few of the battle-hardened veteran cells are retained by the immune system in the event of a subsequent attack by the same virus. These memory cells "remember" specific viruses and respond more quickly when presented with the same foe. This is an essential evolutionary advantage because some viruses have the ability to hide in the body and then resurface when the immune system is temporarily weakened. Herpes simplex—the cold sores some people periodically get on their lips—is an example of one type of virus that is never completely defeated.

After repeated attacks by the same virus, more and more memory T-cells are created, replacing naïve T-cells in the body's immune system. As the years pass, these memory T-cells become weak and less effective, but they take up more and more of the immune system's limited defensive resources. Like old, inept generals who prepare for the next war using outdated strategies, these aged memory cells eventually get in the way of the proper function of the immune system, effectively crowding out many naïve T-cells that would otherwise protect us against new kinds of viruses.

According to some immunologists, one of the major culprits here might be one specific virus—cytomegalovirus (CMV). It's so common that 85 percent of adults over age 40 are infected by it, yet as diseases go, CMV wouldn't rate a second glance as it rarely causes recognizable signs of illness. When it flares up, about half of its victims suffer no symptoms at all, while the other half are afflicted with only minor, hard-to-diagnose symptoms, such as aches and pains or fever. So while it is weak in symptoms, CMV is extremely persistent, attacking over and over throughout our lifetimes. With each new infection, more memory cells are generated to fight CMV and a few of them remain after each battle. Over time, these old CMV memory cells even lose their ability to fight the CMV virus, which causes even more CMV memory cells to be created upon the next attack.

Interestingly, in a casual conversation, one cardiologist suggested that CMV might also be one of the infectious causes of myocarditis— inflammation of the heart. However, it is almost never cited as a cause because the virus flares and ebbs, while the heart damage is permanent and accumulates with each attack. Thus, the CMV inflammation would have occurred, caused the damage, and been gone long before the heart disease was discovered.

Since the blood can only hold a certain number of T-cells, we can't fix this problem by creating more T-cells. The anergic (inactive) memory T-cells must be destroyed while leaving naïve T-cells unharmed. The repair solution would be a "smart bomb" (an antibody or enzyme) to selectively target only the anergic T-cells. Another approach, of course, would be to find a way to kill all viruses completely instead of allowing them to stick around and attack each time the immune system is temporarily weakened.

## EXTRACELLULAR STIFFENING

With age, the tissue of the heart and the major arteries begin to harden. The heart can no longer function as efficiently, and the arteries can no longer stretch as much with each contraction of the heart. It takes a long time for this damage to reach pathological levels, but it is a contributing factor in the majority of deaths over the age of 100. Four of six deaths in one group of seniors over the age of 100 could be attributed to extracellular stiffening of the heart and arteries.

This loss of elasticity of muscle tissue is caused by the cross-linking of two proteins with a glucose (sugar) molecule. The result is somewhat like gluing two rubber bands together. Individually they can stretch, but when glued together side-by-side to form one band—if the glue is strong enough—it hardens the rubber to the point where these bands can't stretch at all.

Muscle fibers are like hundreds of rubber bands tied together end to end, forming a long stretchy strand of tissue. Hundreds of these strands aligned parallel to each other form the muscle itself. When molecules of glucose bind some of these individual rubber bands on one strand with individual bands on an adjacent strand, the two strands can still stretch, but not as much as before. This is because some individual fibers—now cross-linked to fibers on another strand by glucose—won't stretch at all.

As more and more strands of protein are cross-linked together, the muscle tissue loses more and more elasticity. Eventually the stiffening of this tissue can cause systolic heart failure, diastolic heart failure, stroke, diabetic retinopathy, cataracts, hypertension, hardening of the arteries, and arthritis. Diabetes—affecting 20 percent of seniors and increasing due to rising obesity—is a major contributing factor to extracellular stiffening because it floods the blood with excess glucose, greatly increasing the risk of age-related diseases and premature aging.

## MITOCHONDRIAL MUTATIONS

The mitochondria are the energy factories of the cell, converting raw materials from food—mostly glucose and other food molecules—into adenosine triphosphate (ATP), which is the primary energy source for muscles, cells, and the numerous biochemical mechanisms of the body. Each cell contains dozens of mitochondria that are like tiny power plants, converting raw materials into energy, but also belching out dangerous byproducts in the form of free radicals. In the normal course of events, these highly active mitochondria wear out or are damaged by the same free radicals that they produce. Subsequently, old or damaged mitochondria are destroyed by the body's natural garbage disposal systems and are replaced by fresh mitochondria every few weeks.

Some of these damaged mitochondria mutate to such a degree that the body can't identify them for routine destruction. As their numbers grow, they gradually crowd out the healthy mitochondria. These mutated mitochondria create what scientists call reductive hotspots—locations where superoxide free radicals hitch a free ride on passing LDL cholesterol particles, spreading free radical damage throughout the body.

## NUCLEAR MUTATIONS

The nucleus is the blueprint library of the cell and contains our genetic DNA. When the nucleus mutates, it typically loses its ability to replicate and eventually dies, so there is no long-term damage, but in rare occasions, these mutations can replicate themselves, which can lead to cancer. Cancer cells reproduce rapidly without suffering the telomere shortening found in normal cells. As a result, cancer cells are essentially immortal. Unless the body's innate immune system manages to kill cancer cells faster than they can reproduce, cancer can spiral out of control and eventually lead to death.

## THE GATHERING IN THE PARLOR

Since the ancient Greeks first pondered the meaning of life, scientists and philosophers have searched for causes of aging and death. Modern science has seemingly found the answers.

We are no more programmed to die than we are programmed to fly or breathe under water. Instead, evolution favors genetic variants that optimize early survival and reproduction, and once we reproduce and care for the offspring, we are no longer needed for natural selection to take place. And to reiterate the automobile example, the body simply wears out, like an old car that was never built to last forever. Abuse accelerates the aging process, but even if the car is driven normally or just sits, eventually rust and age take their toll. Gaskets wear out, becoming inflexible and brittle (cross-linking). Oil becomes thick with particulates that damage the rings and cylinders of the engine (intracellular and extracellular junk). The edges of the gears in the transmission wear down and no longer function properly (telomere shortening). The radiator becomes clogged with particulates, allowing the temperature to get dangerously high outside its normal operating range (failure of homeostasis).

In an interesting twist of art imitating life, the cause of death itself is surprisingly like the board game *Clue*, in which participants attempt to solve a hypothetical murder, to figure out where that murder took place, and with what weapon. In each game, random chance and circumstance ultimately determine the eventual killer. And so it is in life. There is no one certain factor in death. Instead, the causes of death occur randomly. One thing wears out and then another. Something breaks here; something breaks there. One organ begins to fail and then a second one does. Random chance and circumstance—like exposure to a flu virus or repeated stresses that overwhelm the immune system—all play a contributing role in aging and death.

So has the world's greatest murder mystery been solved? Not quite—after all, these killers are still at large, resulting in the deaths of 100,000 seniors every day. Now that we have a much better understanding of the causes of aging, stopping them should be an international priority. Fortunately, we don't need to know the causes of aging with certainty to combat its effects. Even more fortuitously, solutions to repair the damage done by aging are much closer than you might think.

# SIX

# Repairing Damage and Extending Work Span

In the previous chapter, we discussed one of the great mysteries of all time—the causes of aging. However, we now find ourselves facing another question about aging that promises to be far more daunting: what can be done about it?

Aging is perhaps the most universal problem facing mankind. Granted, a myriad of serious global problems surrounds us, but they don't affect each and every one of us. Hunger for example, is a major problem in underdeveloped nations, but for most of the developed world, the threat of hunger has been resolved. The same can be said for malaria, tuberculosis, and malnutrition—all these problems continue to exist, but they no longer significantly impact developed nations. AIDS and HIV remain a major crisis in Africa and parts of Asia, but these diseases are now largely under control in the developed world. Even global climate change—widely discussed in the most apocalyptic of terms—might arguably benefit some people in cold climates, whereas aging and death affect everyone.

The issue impacts even those who are not old because social programs for seniors will be largely funded by taxes on younger workers. Throughout the world, the cost of senior social programs, such as Medicare, is becoming increasingly unsustainable. Unchecked, these programs will place an enormous financial burden on today's youth. Although the financial liabilities of Social Security and Medicare have been widely reported, these problems are much worse than most people realize, as you'll learn in chapter 8.

If the damage caused by aging could be postponed or even reversed by a few years, the financial benefits could be immense. Extending work span—the maximum number of years that workers can remain productive—by just five years could postpone most of the financial problems of Social Security and Medicaid for years. The monies these senior workers put *into* the system—instead of being taken out—would provide a grace period for researchers to discover more ways to extend senior health span, and thus work span, even further into the future.

Today, the age of full Social Security eligibility is 66. If it were to be extended to age 71, the savings in Social Security revenue alone would top $180 billion per year.[1] The actual benefit to governments could be even greater since these working seniors would continue to pay federal income taxes and Social Security withholding tax on their earnings, generating an additional $60 billion per year. (That assumes, of course, that jobs would be available for seniors and that these seniors would not displace other workers.)

We'll discuss these options in chapter 8 of this book, but the magnitude of savings clearly illustrates why policymakers have a strong financial incentive to fund research into repairing and slowing age-related damage. A few tens of millions of dollars prudently invested in promising areas of research to repair age-related damage could literally return hundreds of billions of dollars in the future.

Considering such a spectacular return on investment, government funding of research to repair age-related damage and slow the aging process is not just prudent, it would be financially irresponsible not to do so. The question is no longer whether we should fund such research, but rather what types of research should be funded. Logically, we need to turn our attention first to the low-hanging fruit—that is, the issues that are easiest to repair and most likely to affect the productivity of young seniors. Examples include better hormone replacement therapies for men as well as women, better obesity treatments, and research on how best to improve the fitness of senior workers. Next would come repairing damage due to diseases and old age and, finally, finding cures for major diseases and aging itself.

## MISCONCEPTIONS ABOUT SENIOR WORKERS

In order for seniors to keep working past age 65 in large numbers, several major obstacles need to be overcome—each presenting its own unique challenges. While the primary focus of this book is on the biological challenges of tissue and cellular repair, we can't ignore the social and cultural obstacles that must also be overcome to maximize human work span.

Perhaps the biggest nonbiological challenge is the perception that older workers are less productive and cost employers more. A survey by the Wharton Center for Human Resources found that 49 percent of employers felt older workers did not keep up with technology; 38 percent believed older workers led to rising health-care plan costs.[2] Some employers complained that older workers just "retired in place," doing the bare minimum to get by. One employer felt older workers lacked a strong work ethic, pointing out that all his older workers left work for the day before the younger workers.[3] Given equal abilities, it's obvious that an employer would prefer a younger worker who puts in 65 hours each week versus an older worker who only puts in 40 hours a week. Other employers complained that seniors expected to be paid more even when doing the same job as younger workers who, by comparison, were just glad to have a job and the opportunity to prove themselves.[4] Older workers were also cited as being picky about health-care benefits, which is less of a concern for young workers.[5]

Some employers felt today's rapidly changing work environment presented more of a problem for older workers. Besides lacking the latest technological skills, seniors were perceived as lacking the desire to learn them. Another knock against older workers was the belief that they have difficulty adapting to change—the "We did it this way at XYZ Corporation" attitude. One employer in a fast-moving technology industry proclaimed that "fail fast" is the new industry mantra.[6] Make a quick decision and if it turns out to be wrong, immediately change course. Meanwhile, older workers are still ingrained in the old "get-it-right-the-first-time" approach to decision making. What older workers viewed as quality control, this fast-moving employer viewed as a lack of aggressiveness or caring about the job.

Employers also pointed out that older workers could be more disruptive because they often reported to much younger managers. What the older workers apparently viewed as efficiency or avoiding mistakes, these younger managers interpreted as a know-it-all attitude.[7] This not only interfered with the older workers' productivity, it also hindered the productivity of the entire team.

All of these beliefs are, of course, stereotypes, but that does not make them any less of an obstacle for aging workers or today's employers who will be hiring seniors in the foreseeable future. They see seniors as presenting a higher risk for such behaviors, and when faced with the opportunity of hiring an older versus a younger worker, they opt for the lower-risk scenario.

## THE NEW OLDER WORKER

These stereotypes, however, come from a subset of older workers who are now retired. Baby boomers are the next generation of older workers, and they are quite different from the previous generation. Boomers don't see themselves as old—quite the opposite. This new generation of older workers promises to be quite different from their parents. For instance, boomers are rejecting the traditional retirement community of their parents for newer intergenerational communities. They not only reject the premise of getting old, but they also clearly show no hesitation in spending sizable amounts of their discretionary income to stay as young as possible for as long as possible. Still, boomers are a very large demographic.

As they approach retirement age, those who stay in the workforce will fall into four distinct categories.

First, some boomers enjoy what they do and plan to continue working well past normal retirement age. These workers gain a greater sense of fulfillment from their vocations than they would from retirement. In fact, most of these seniors insist they would much rather continue to work than retire. Not surprisingly, these seniors are very good at what they do and are highly valued by their employers, so much so that a common lament is, "I don't know what we'll do" when the employee retires. Also not surprisingly, these seniors tend to be reasonably healthy or have artfully adapted their responsibilities at work around their physical limitations.

The second category consists of workers whose lives revolve around their work. These workaholics often stay late at their jobs and are reasonably good at what they do. Unlike the fulfilled workers in the first category, however, they tend to be in comparatively poor health because health is just another aspect of life they've neglected due to their obsessive focus on work. While this may not be a good outlook to have, one positive characteristic, as far as work span is concerned, is that members of this category—much like the fulfilled workers—will keep working as long as their health allows them to do so.

The third category contains workers who unexpectedly find it necessary to work a few more years before they can afford to retire. In some cases these workers were counting on using their home equity as a retirement fund, only to see their homes' values drop precipitously after the housing bubble burst in 2007. Others lost a substantial portion of their retirement funds during the stock market decline of 2008. There are likely many other reasons why these seniors may need to work a few more years before retiring—perhaps an untimely layoff, the expense of putting children through college, helping adult children who fell on hard times in a struggling economy, or finding that they simply failed to save enough to retire comfortably.

Seniors in the final category have little or no savings of any kind. They failed to plan for retirement and have no choice but to keep working.

Although the underlying personalities of each of these four categories of boomer workers are quite different, they have something in common. None of them are likely to retire in place, doing just the minimum to get by. They either enjoy working or they desperately need to maintain employment. Plus, tomorrow's older worker will be younger in mind and attitude than previous generations. This is a new ball game with new rules, and while boomers have not shown themselves to be the smartest of savers, it's fair to say the majority have learned how to play the game. If the new rules say they need to learn the latest technology to maintain job security, they'll do it. If they need to say, "Yes, sir," to a supervisor half their age, they'll do that as well.

## EMBRACING THE NEW OLDER WORKER

In addition to the changing personality of the senior worker, there's a groundswell of employers who embrace the idea of an older workforce. Recent studies show older workers use fewer sick days and experience lower absenteeism rates than younger workers. In spite of the concern over older workers raising the costs of health-care plans, health care for older workers often costs less because they no longer need to include children on their plans. Once older workers reach age 65, they become eligible for Medicare, further reducing employers' cost of health insurance. Older workers typically possess better interpersonal skills and better customer service skills than their younger counterparts. They're more patient, more mature, more dedicated, and more amenable to working part time. The new generation of seniors will work hard, adapt to new technology, and be able to make quick decisions.

In addition to adjusting to younger employees, seniors must display some other major attitude adjustments to flourish in the new senior-friendly workplace. Some of today's workers believe they're entitled to retire at age 65—overlooking the fact that full Social Security benefits don't start until age 66 (or in the case of France, age 60). As far as they're concerned, they paid into the system and they are entitled to what they paid for. So these seniors themselves may display a tremendous resistance to changing the retirement age. Nonetheless, retirement age in the United States (as well as in France, Great Britain, and the rest of the European Union) eventually must change because it would be fiscally irresponsible not to do so.

Cultural norms, like retirement age, stubbornly resist all efforts to change. In the mid-1980s, then–Speaker of the House Tip O'Neill allegedly quipped, "Social Security is the third rail of politics."[8] Touch it and you die—or words to that effect. For decades, O'Neill proved to be quite prescient as attempts to reform Social Security by every subsequent administration failed.

On the other hand, some stubbornly held cultural norms have changed radically over relatively short periods of time. In 2006, mainstream articles indicated the majority of baby boomers hoped to retire by age 62 or earlier. Indeed, at that time, the age at which most workers retired had been gradually trending downward for a couple of decades. Then the housing bubble burst and the stock market crashed, leading to the highest unemployment levels since the Great Depression. By 2010, a remarkably short time span in terms of cultural change, the expectations for retirement had completely reversed. Surveys in 2010 showed that the typical worker approaching retirement age planned to retire two to four years later, while the majority indicated they would work past age 65.

The resulting high levels of unemployment and the depressed economy have accelerated the cultural and societal changes necessary to make the workplace more senior friendly. To be sure, the psychological issues surrounding what seniors have been promised or feel entitled to will be challenged in the future, but those issues can be addressed with well-planned individual and business incentives. Once these psychological and cultural issues are addressed, one major and seemingly insurmountable obstacle still remains—the risk of declining productivity due to aging and age-related health problems.

## HEALTH BARRIERS FOR OLDER WORKERS

As we discussed earlier in this chapter, the ideal approach would be to re-pair damage that affects the productivity of relatively young seniors first, adding 5 to 10 years to health span (and thus to work span). Damage that doesn't become pathological until age 75 or 80 could be addressed later. In effect, this approach would keep younger seniors healthy enough to participate in the workforce for another decade, at which point another decade of medical advances would have transpired, potentially increasing the work span even more.

Surprisingly, the health challenges facing younger seniors are not pri-marily major diseases, such as cancer and heart disease. Although many workers reaching age 65 will have one or more chronic diseases, these conditions only hinder job performance in the most severe cases. Instead, the first level of age-related damage to affect productivity might be more accurately referred to as measures of well-being, that is, health issues that aren't necessarily disease-specific.

The first of these is decreased stamina. A common complaint of older workers approaching retirement is that they've just grown increasingly tired of the day-to-day demands at work. Chronic health conditions can play a part in this weariness, but it's often present in those without major diseases, which is why it is more appropriately referred to as a measure of well-being. Most jobs aren't particularly physical—it's just the daily grind and stress that wears workers down. Of course, this applies to average jobs, not extremely physical occupations like firefighters, construction workers, or manual laborers. It's not uncommon for some workers to no-tice a decrease in stamina as soon as age 40, but for most, it starts around age 60.

Decreasing energy is closely related to a lack of stamina and typically follows it by a few years. It can, however, occur suddenly when the cause is disease-related—for example, following recovery from a stroke or heart attack. While a lack of stamina tends to impact workers toward the end of the workday, a lack of energy potentially hinders performance from the moment the worker shows up at the job. A noticeable decrease in energy level normally begins to surface around age 60, but it can appear earlier in the presence of age-related diseases or other factors we will discuss later in this chapter.

A third characteristic that could potentially impair the productivity of older workers is decreasing enthusiasm. It's not necessarily a lack of enthusiasm about the job itself (although that may certainly be the case), but more often a lack of enthusiasm about life in general. Of course, whether a worker loves a job or hates it will contribute to the overall level of enthusiasm, but it can also be triggered by physical conditions, such as declining hormone levels or the side effects of medications used to treat pain or chronic medical conditions. Regardless, there is no doubt that enthusiasm can go a long way to overcome other employee shortcomings. In fact, if employers were forced to pick one nonfinancial reason for preferring younger employees over older ones, it would probably be that younger employees bring enthusiasm to the workplace. Energy and enthusiasm are infectious—and so, frankly, is the lack of them.

While other conditions affect the workplace productivity of older workers, there is no doubt that if the stamina, energy, and enthusiasm of an older worker matched that of a 35-year-old, it would do a great deal to level the playing field between younger and older workers. Unfortunately for some older workers, these key productivity variables have already begun to decline. It could be due to an age-related disease, but often there is no single cause. Perhaps the daily grind has simply taken its toll. For a while, the wealth of experience these older employees bring to the workplace may compensate, but eventually one or more of the key productivity factors will decrease to the point that it affects their performance. This not only affects the employer, but it also reinforces the stereotype of older workers as poor employees, which in turn makes it more difficult for other older workers to get hired in the future. At the extreme end, this perpetuates age discrimination and could force the government to step in and pass well-intended laws—laws that sometimes have unintended consequences that run counter to their intended purpose.

## THE AGE DISCRIMINATION IN EMPLOYMENT ACT

The Age Discrimination in Employment Act of 1967 (ADEA) is a good example of a law that is actually counterproductive to its intent. The ADEA made it unlawful to discriminate based on age in the hiring, firing, laying off, compensating, distributing benefits, assigning jobs, and training of job applicants and employees. There are many reasons why a worker's productivity might decline other than age, not the least of which may be that the employee is an unsatisfactory worker. Regrettably, the ADEA's vague guidelines made it possible for some of these ineffective workers to sue successfully on the basis of age even though their substandard performance had been thoroughly and properly documented. It's no surprise that employers began to view the potential for litigation as a major risk in hiring senior employees. In their minds, the way to reduce this risk of unfounded litigation was obvious—don't hire older workers. Ironically, the very law intended to make the workplace more senior friendly instead made it harder for older workers to find jobs.

Fortunately, in 2008, the Supreme Court clarified this law. This ruling now requires the employee to present direct or circumstantial evidence demonstrating improper action by the employer. Hopefully, this will allow employers to hire older workers without fear of future lawsuits should they need to fire an employee for failure to meet the same performance standards as younger employees.

## COPING WITH CHRONIC CONDITIONS

Eventually, many older workers begin to experience age-related medical problems, but often these problems can be treated without a decrease in job performance. For example, it's not uncommon for workers to perform in extremely high-level, stressful jobs in spite of heart problems. Dick Cheney, for example, has serious heart problems, yet he served for eight years as the Vice President of the United States. Cancer is another major health issue, but it does not always result in a decline in productivity.

# Chronic Conditions Affecting U.S. Adults 70+

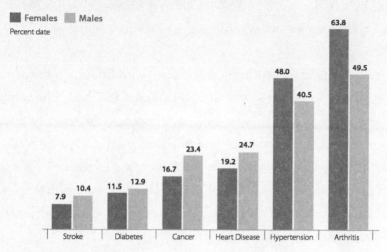

Source: William D. Novelli, AARP, Healthy Aging: Preventing Disease and Improving Quality of Life Among Older Americans 2003, CDC, 2011.

*Figure 6.1. Chronic conditions affecting U.S. adults over 70.*

Although chemotherapy can reduce energy and stamina, once it is completed, cancer is not usually an ongoing factor in workplace performance.

After age 70, however, age-related disease does become a bigger factor for the typical senior worker. Age-related diseases that can affect performance move to the forefront, notably high blood pressure, cardiovascular disease, high cholesterol, high blood sugar (diabetes), and arthritis. Fortunately, except in extreme cases, these conditions can be effectively managed. There are, however, two other very common conditions that can potentially hinder health and productivity for seniors: declining hormone levels and obesity.

## DECLINING HORMONE LEVELS

Female hormone levels decline dramatically after menopause, but the hormone decline in males is equally dramatic, albeit over a much longer time frame. By age 65, male testosterone levels have dropped by 50 percent compared to their peak in young adulthood. Low hormone levels adversely impact many attributes of a productive employee—enthusiasm, strength, stamina, emotional equilibrium, and ability to cope with stress, just to name a few.

Age-related decline in hormone levels can be profoundly influenced by a healthy lifestyle, which we will discuss in more detail later. Hormone replacement therapy (HRT) goes beyond lifestyle to increase lagging hormone levels with supplemental hormones, medications, or hormone boosters. Sadly, it doesn't appear that HRT increases lifespan or life expectancy, but the results are promising that it does increase health span—that is, the number of years that one can remain relatively healthy, and thus productive in the workforce.

## OBESITY

Most of the age-related changes in healthy 65-year-olds can be addressed with current levels of medical technology. Nutrition, exercise, weight management, and hormone replacement therapy are all sufficiently advanced to help keep seniors healthy well into their 70s. Unfortunately, most workers are not reaching age 65 in good health due to poor lifestyle decisions, such as poor nutrition or lack of exercise.

So far we haven't discussed the single greatest issue that aging boomers must overcome if they want to maintain good health and careers past age 65: obesity. In the United States, 35 percent of adults are obese according to the Centers for Disease Control and Prevention, and the percentages are even a couple of points higher for baby boomers.[9] Obesity greatly limits stamina and energy. It inhibits the production of good hormones and greatly increases the risk of debilitating and painful diseases, which often necessitates the use of painkillers which come with related potential side effects. Obesity increases the risk of arthritis, joint pain, back pain, neuropathic muscle pain, and especially diabetes, which is over twice as prevalent in the overweight population and up to several times more prevalent in obese populations.[10] Since more than 20 percent of seniors are currently diabetic, this is an ominous warning for baby boomers, who struggle with obesity at a higher percentage than the current generation of seniors. Diabetes also dramatically increases the risk of cancer and even Alzheimer's disease.[11] One Swedish study found that obesity combined with high blood pressure and high cholesterol (both fairly common in the obese population) increased the risk of Alzheimer's disease by 500 percent.[12] Some may argue whether or not obesity is a disease, but there is no question that it needs to be addressed as an aging issue, because it accelerates age-related damage.

Because obesity greatly increases the risk of Alzheimer's disease and other debilitating illnesses, it also increases the potential need for assisted living. The financial burden that assisted living could in the future place on states and the federal government is breathtakingly large. The cost of one year in a long-term care facility (currently averaging $79,000 per year) plus the cost of medical care and prescription drugs could easily top $100,000 annually. The cost of care for 1 million patients could top $1 trillion. That is not an unlikely scenario—it is, in fact, quite probable as more seniors live to advanced age and the risk of Alzheimer's disease increases to about 50 percent by age 85.

### RESEARCH PRIORITIES

So far, we have identified two areas of concern that must be given top priority. The first includes productivity characteristics such as stamina, energy, and enthusiasm, which are ultimately what employers look for in their workforce. The second encompasses a series of factors that adversely influences these characteristics, such as pain and the side effects of pain-killers, arthritis, decreasing hormone levels, and obesity.

In terms of repairing damage, none of these early productivity-limiting conditions fall neatly into one of de Grey's seven classes of damage, so it's difficult to build a convincing argument that one class of damage repair should be addressed before the other six. What first causes cellular or tissue damage is a function of individual genetics, lifestyle, and environment. For each individual, these variables will be different and even subject to random chance. Some individuals will be impacted by cell mutation (cancer) while others will not. Some will develop diabetes—exacerbated by mitochondrial malfunction—while others will not. Some will develop dementia or Alzheimer's disease, while others will stay mentally sharp into their 90s. As a result, it's unrealistic to suggest that one type of damage repair research should receive priority over another.

Fortunately, there is really no reason why simultaneous research could not go forward in multiple areas of repairing damage. A breakthrough in one area might have medical implications for another, as has often been the case in medical research. In fact, there are two research areas that have already shown great promise for repairing several classes of damage and, ironically, have been underfunded to date—regenerative medicine and stem cell research.

# SEVEN

# Recent Advances in Biogerontology and Regenerative Medicine

Advances in regenerative medicine over the past ten years have been breathtaking in scope, yet they have largely slipped under the radar of mainstream awareness. Even some professional researchers aren't aware of the implications of these discoveries. Fortunately, I've seen this progress unfold firsthand.

Ten years ago when I finished my graduate work at Johns Hopkins and the Moscow State University, the future of aging research was very uncertain, but it seemed to hold great promise. So instead of pursuing a specific line of research after my graduate studies, I went to work for a company called GTCbio as Director of Business Development. GTCbio specializes in organizing high-level biotechnology meetings worldwide, bringing together the top minds from academia and industry. While attending these conferences, I had the exceptional opportunity to talk with thought leaders and learn the state of the art in a wide range of medical and technological research: cancer, biomarkers, diagnostics, stem cells, immunology, metabolic diseases, oxidative stress, apoptosis, and drug discovery. Often, I would become aware of research results months before they were published or breakthroughs years before clinical trials were officially completed. Over the years, I became friends with many scientists throughout these fields, and I'm even involved in their work in one way or another. Some of these scientists have joined the science advisory board of the International Aging Research Portfolio project that I founded in 2010.

## WHAT IS REGENERATIVE MEDICINE?

The definition of regenerative medicine is very broad. The National Center for Biotechnology Information defines it as "repair or replacement of damaged, diseased, or metabolically deficient organs, tissues, and cells via tissue engineering, cell transplantation, artificial organs, and bioartificial organs and tissues."[1] It brings together such diverse fields as nanotechnology, lasers, stem cell research, and tissue engineering with a common goal of re-engineering damaged tissue back to its original healthy condition.

Regenerative medicine holds exciting possibilities that are almost within our reach, the likes of which you might find in the pages of science fiction novels. Crushed spinal cords could be regrown, allowing paraplegics to walk again. Damaged hearts could be repaired just by replacing damaged tissue with new healthy tissue grown from patients' own adult stem cells. Entire organs could be grown in laboratories and then safely implanted into the patients' own bodies, eliminating the need for organ donors and the tragedy of patients dying while languishing on the waiting list for a new heart, kidney, or other major organ.

## STEM CELL RESEARCH

Stem cells and stem cell research are widely misunderstood by the general public and have become the subjects of a very polarized political debate. As a result, large segments of the public are misinformed when it comes to stem cell research. First, there is still a misconception that there is only one type of stem cell and that it can only be obtained from a fetus. In reality, there are several different types of stem cells—embryonic stem cells

(ESCs), adult stem cells, and induced pluripotent stem cells (iPSCs), just to name a few—and each will be discussed in more detail in this chapter. Second, there's a widespread misconception that embryonic stem cells must be obtained from an abortion. As you'll soon learn, that is not the case at all.

Finally, among those who are knowledgeable in the first two areas, there is a misconception that the discovery of induced pluripotent stem cells and transdifferentiation—the ability to convert one type of cell directly into another type of cell without going through the iPSC stage—renders experimentation with embryonic stem cells completely unnecessary. That is also a fallacy. Taken together, these misconceptions have significantly hindered the advancement of stem cell research.

Interestingly, stem cell therapy was actually used on humans before stem cells were known to exist. Several bone marrow transplants were successfully performed in France after a radiation accident in the late 1950s, but the exact reason why the patients recovered was unknown. Then in 1961, Dr. James Till and Dr. Ernest McCulloch published research proving the existence of stem cells. Stem cells are unspecialized cells that have the remarkable ability to transform into other types of cells. A fetus, for example, shortly after conception is almost entirely unspecialized stem cells. In a matter of days, these cells begin to divide, creating all the different types of cells found in the body—bone, tendons, muscles, organs, and so on. However, some types of stem cells continue to exist even after we reach adulthood. Dr. Till and Dr. McCulloch found that stem cells transplanted into bone marrow had apparently revitalized the damaged tissue and allowed it to return to its normal function of producing new blood cells.

Little did anyone realize at the time that stem cell research would become one of the most highly politicized medical research topics of all time.

In 1974, Congress banned federally funded human fetal tissue research until guidelines could be established. In 1975, the Ethics Advisory Board for fetal tissue research was established by Congress, but this board didn't survive for long. In 1980 President Reagan allowed the board to expire without action, which in turn caused a halt to federal funding for human embryo research. In 1990, Congress attempted to override this moratorium, but President George H. W. Bush vetoed the measure. In 1993, President Bill Clinton lifted the moratorium by executive order, but when he received thousands of letters urging him to reverse his position, he did so, once more canceling federal funding for human embryonic research. In 1995, Congress took this ban one step further, passing the Dickey-Wicker Amendment, which prohibited the use of federal funds for the creation of human embryos for research purposes or research in which human embryos are destroyed.[2]

Despite the lack of federal funding, privately funded stem cell research continued, albeit at a much reduced pace. This is mainly because corporations have been historically reticent about allocating funds until government-funded research first shows that the project has commercial potential. In 1998, in a landmark discovery that some consider to be the real beginning of Regenerative Medicine 1.0, University of Wisconsin scientist James Thomson isolated human embryonic stem cells and proved their ability to rejuvenate old or damaged tissue.[3] Ironically, his discovery only exacerbated the political debate, because his team obtained stem cells through a process that destroyed human embryos.

On August 25, 2000, the NIH published guidelines for federal funding of human embryonic stem cells. Among these guidelines were stipulations that the cells must be derived with private funds using frozen embryos from fertility clinics. Further, there must be more embryos than the donor needs and obtained with the consent of the donor. The guidelines lasted for less than a year. In August 2001, only eight months after taking office, President George W. Bush prohibited federal funding of any research using human embryonic stem cell lines derived after August 9, 2001. However, he allowed for federal funding of stem cell research using human embryonic stem cell lines registered before that date. In 2007, President Bush signed an executive order that slightly expanded the number of human stem cell lines that can be used while keeping the previous ban in place. In March 2009, President Barack Obama rescinded that executive order, removing barriers to federally funded human stem cell research. But in August 2010, a federal judge blocked his order, saying it violated a ban on federal funds being used to destroy human embryos. This ruling was itself later overturned.

And so it goes—medical research has become so politicized that corporations are hesitant to pursue human stem cell research in the United States regardless of the favorability of any given current law because those laws have the potential to be reversed in the future. Fortunately, in 2007 a breakthrough discovery occurred that should have changed the nature of this debate forever—the creation of a brand new type of stem cell called induced pluripotent stem cells.

## INDUCED PLURIPOTENT STEM CELLS

In 2007, Shinya Yamanaka of Kyoto University announced he had successfully reprogrammed adult mouse cells to behave like embryonic stem cells. His team called these reprogrammed cells induced pluripotent stem cells or "iPSCs." About the same time, James Thomson and his University of Wisconsin coworkers reported that they had also created human-induced pluripotent stem cells using a somewhat similar approach. Yamanaka's and Thomson's breakthroughs appeared to leap over the main obstacles scientists faced by using embryonic stem cells. From a purely scientific perspective, iPSCs also seemingly eliminated the problem of immune rejection because cells from the patient's own body could be used. Thus, the body's immune system would not identify the cells as invaders and try to destroy them.

Adding to the excitement of Yamanaka's and Thomson's discovery was the simplicity of the process. "It's really easy—a high school lab can do it," opined Mahendra Rao, head of stem cell and regenerative medicine at Invitrogen Corporation in an August 2008 *Scientific American* article.[4] Rao's remark might be a bit of a stretch, but there is no doubt that creating iPSCs had been made far simpler. A 2012 PubMed search for "induced pluripotent stem cells" produced hundreds of articles. Since many scientists don't publish until their research is complete (and some corporate scientists don't publish at all), the actual number of iPSC research projects could be in the thousands.

Since iPSCs don't require the harvesting of human embryonic stem cells, the major ethical stumbling block for stem cell research had seemingly been eliminated. Sadly, scientists and politicians with a pro-life agenda argued that the discovery of iPSCs made the need for human embryonic stem cell research completely unnecessary. In reality, there are things that iPSCs can't do that embryonic stem cells can do. Researchers might be able to find solutions to these limitations in a few years, but meanwhile important stem cell research could be continued using embryonic stems cells. At some point in the future, iPSCs might improve to the point that ESCs would no longer be necessary.

Meanwhile, large portions of the general public still believe that *all* stem cell research begins with embryonic stem cells. As a result, they are against *all* types of stem cell research. Even today when I discuss stem cell research with individuals who are not scientists, they are often surprised to discover that stem cells can be obtained without abortions.

Still, the discovery of iPSCs revolutionized the world of stem cell research, so much so that Yamanaka won the 2012 Nobel Prize in Physiology or Medicine for his discovery.[5] New research on iPSCs is proceeding at a breakneck pace. Hardly a week goes by without some new research being published on clinical applications ranging from autologous stem cell transplantations (using cells from one's own body) to artificial organs made with iPSCs.

In 2011, University of Wisconsin researchers announced that they had created iPSCs from the bone marrow cells of a leukemia patient. This is significant because abundant supplies of bone marrow already exist in blood banks, which makes iPSCs more accessible. Equally important, it was the first time that scientists had successfully reprogrammed iPSCs from diseased blood cells, providing a ready source of stem cells that carry the markers for specific diseases, such as leukemia. By using iP-SCs from leukemia patients, researchers will be able to create leukemia *ex vivo* (outside the body) to test experimental leukemia drugs and therapies on human tissue without endangering human patients.

As exciting as the future appears to be for induced pluripotent stem cells, a new type of man-made stem cell offers even more promise for the future.

In 2011, scientists at Johns Hopkins developed a process to create stem cells nonvirally by using plasmids—small DNA molecules—to induce the cells to become pluripotent stem cells.[6] Previously, viruses were used to carry genes into the cells to cause them to transform into iPSCs. Viruses can mutate, so the virus-free approach was a leap forward in stem cell research because it eliminated the risk that the virus used to create the stem cell might eventually mutate, causing cancer or triggering the body's immune system to reject the tissue. The nonviral approach to create stem cells is also far less expensive, costing only one-tenth that of virally-induced iPSCs.[7]

## TRANSDIFFERENTIATION

Once Yamanaka successfully created iPSCs in 2007, researchers were encouraged to rethink their entire approach to stem cell research. If an adult cell could be induced to dedifferentiate backward to its pluripotent state and then redifferentiate forward to a different type of adult cell, might it even be possible to bypass these steps? Could one type of adult cell—skin cells for example—be converted directly into heart or other organ tissue without first being reversed into pluripotent cells?

Since all cells in the body contain the same genes, scientists agreed in principle that it was possible. The difference between a skin cell and a heart muscle cell is that different sets of genes are silenced in each different cell type. If scientists could induce the genes within a skin cell—or any other cell type for that matter—to mimic the gene expression pattern of a heart muscle cell, the skin cell would theoretically become heart muscle without going through the dedifferentiation and redifferentiation steps required when using iPSCs.

Encouraged by Yamanaka's success, scientists began looking at direct conversion of cells—called transdifferentiation—as the next step in regenerative medicine. One of the first successes was the conversion of skin cells into heart muscle cells by Deepak Srivastava, director of the Gladstone Institute of Cardiovascular Disease in San Francisco. He found that activating three genes was sufficient to convert a type of cell found in skin tissue into heart muscle cells. When these cells were implanted into mouse hearts, they worked just like normal heart muscle cells. Later, the same researchers replicated hair and eyes using the same technique. Not long thereafter, a damaged liver was repaired using connective tissue cells taken from the tip of mouse tails, providing proof that cells reprogrammed through transdifferentiation could repair damaged organs.[8] In 2011, cell biologist Lijian Hui of the Shanghai Institute for Biological Sciences in China found that expression of three proteins and suppression of one protein was sufficient to convert fibroblasts from the mouse tails into liver cells.[9]

The successes to date prove that transdifferentiation works and has great potential for future therapy development. While the process may be complex, the approach is straightforward: determine what genes need to be expressed or silenced to create the desired type of cell, determine what existing cells are closest to that gene pattern, and then program those genes to act in the appropriate manner to create the desired cell.

Will transdifferentiation eventually replace stem cells as the method of choice for future regenerative medicine applications? It's hard to say. This is a brand-new area of science, so its future is still very uncertain. Five years ago, the exciting new frontier was embryonic stem cell research. Today, it has been joined by not one but three exciting new approaches— iPSCs, nonviral iPSCs, and transdifferentiation. New iPSC discoveries are seemingly being made every week, increasing the probability that major organ repair and even cloning will be possible using some type of stem cell technology in the foreseeable future.

## ARTIFICIAL ORGANS

In 1954, the first successful kidney transplant was completed by Dr. Joseph Murray and Dr. J. Hartwell Harrison in Boston using a kidney obtained from the patient's identical twin.[10] Of course, this advance was of little significance to the average patient without an identical twin. In a more practical surgery, the same team performed the first kidney transplant from a deceased donor in 1962.[11] Transplants of other organs followed rapidly in the 1960s: the first lung transplant in 1963; the first pancreas/kidney transplant in 1966; the first liver transplant in 1967; and the first successful heart transplant, performed by Dr. Christian Barnard in Cape Town, South Africa, in 1967.[12]

In 1984, the National Organ Transplant Act established a nationwide computer registry for organ donors, but obtaining enough donor organs remains a major limiting factor in organ transplants. In the first eight months of 2012 in the United States, about 19,000 kidney transplants—the number-one type of organ transplant—were performed, but 116,000 patients were on the waiting list.[13] A similar trend can be seen for other transplants—each year more people are added to the waiting list than the number of transplants performed. Some people on the waiting list die before their turn arrives, or they are rendered inoperable due to other medical conditions. It's a dire situation for those with failing organs.

The numbers for heart transplants are even more telling. In 2010, there were 2,333 heart transplants, while 3,476 people were on the waiting list.[14] The average wait for a new heart was 133 days. Meanwhile, about 600,000 people die each year due to heart disease, according to the Centers for Disease Control and Prevention.[15] Obviously, not everyone would be a candidate for heart transplant, but it is equally obvious that some deaths could be avoided if more donor hearts were available.

Scientists, looking for a better solution, came up with a novel approach—replace the heart with an artificial pump. The first artificial heart transplant was performed in 1969 by Denton Cooley at the Texas Heart Institute in Houston. It was intended as a temporary measure until a donor heart could be obtained. Unfortunately, the patient died 32 hours later from pneumonia. The first successful artificial heart transplant intended as a permanent replacement was performed in 1982 by William DeVries of the University of Utah in Salt Lake City.[16] The patient who received the Jarvik–7 artificial heart lived for 112 days. The second patient to survive this procedure lived 620 days but with numerous debilitating setbacks along the way.[17]

However, since the number of patients needing heart transplants far outstripped donor hearts, scientists responded by replacing only the section of the heart that was damaged with a mechanical device. One such device, the Jarvik 2000 Flowmaker, is a ventricular assist device, which uses a spinning rotor to boost the flow of blood from the left ventricle to the aorta. It works like a booster pump to increase the volume of blood moved by the heart. As such, it has been far more successful than the earlier complete artificial hearts. The first patient given the Jarvik 2000 in June 2000 returned to a normal life and lived another 7.5 years. When he finally died from acute renal failure in December 2007, his heart was still healthy.[18]

The first successfully implanted self-contained artificial heart completely replaced the existing heart in a patient in 2001. This device, the Abiocor artificial heart, weighs about 2 pounds and utilizes an external battery pack. Its parent company, Abiomed, believes more than 100,000 people in the United States alone could benefit from this device each year. Interestingly, the device's external power supply is not physically connected to its external and internal batteries. Instead, it uses a transcutaneous energy transmission device, meaning no wires or tubes penetrate the skin during battery recharge. Instead, a primary coil is worn externally over an implanted internal coil, thus reducing the risk of infection.

Today, most mechanical hearts still have significant limitations—often the patient has to stay in the hospital in order to be connected by tubes to various external monitors or is otherwise limited in their mobility. As a result, outcomes have not been too promising. As of 2009, the longest time a patient had survived with an Abiocor heart was 17 months.[19]

Companies like Thoratec Corporation hope to improve that survival rate by creating fully portable mechanical hearts and heart-assist pumps. The Heartmate II is a wearable, portable, left ventricle assist device that allows the recipient to return to many normal behaviors. The tiny device weighs only 400 grams and is implanted near the heart, but the batteries are worn in an external shoulder holster, allowing the user to walk and perform other daily activities.[20] The company has several other devices that are already fully operational, such as the Thoratec CentriMag blood pump to provide short-term circulation while other long-term options are being considered.[21]

It's even possible that future devices will no longer attempt to mimic the human heart. In 2011, scientists at the Texas Heart Institute replaced a 55-year-old man's failing heart with two tiny turbines that provide a steady flow of blood without the pulsing normally associated with a beating heart.[22] Since turbines are far less complex and contain fewer moving parts than pulsating mechanical pumps, perhaps they will be the breakthrough that makes mechanical hearts a viable long-term alternative to a donor heart. If mechanical heart turbines can be perfected and proven in clinical trials, these devices could revolutionize the treatment of heart disease.

## BIOENGINEERED ORGANS

As exciting as the future seems for artificial organs, it's possible that bioengineered organs, such as fully functional hearts, livers, pancreases, lungs, and kidneys, could eventually relegate mechanical hearts into obsolescence. No batteries, no bulky external devices—bioengineered organs would be even better than transplants because the risks of rejection by the immune system would be eliminated.

The first step—coaxing stem cells to grow into the appropriate organ tissue—has already been done by scientists numerous times. They've produced beating-heart tissue, liver tissue, pancreas tissue, and more. Since these stem cells must be nurtured to grow *ex vivo*, sophisticated devices called bioreactors are required to provide the necessary nutrients and oxygen as the tissue grows. The science of bioreactors will need to progress in pace with the other areas discussed here in order to achieve widespread application of bioengineered organs.

Another requirement is vascularization—creating the tiny network of capillaries to carry nutrients and oxygen to the organ. Without this network of tiny blood vessels, the bioengineered organ would quickly die. Scientists around the world are pursuing this quest, but so far success has been elusive.

Finally, the organ must be induced to grow into the appropriate size and shape. Scientists have approached this challenge from two fronts. The first method is to create a scaffolding of porous material—made from a variety of polymers or from collagen stripped from a cadaver or porcine organ—that will support the stem cells as they grow over it. In 2009, scientists from Stanford and New York University Langone Medical Center reported the successful extraction of scaffolding material from mice and used it to grow a variety of adult cells, including blood, fat, and bone marrow.[23] This provided proof of concept that stem cells could be made to grow in all three dimensions to create an organ.

Another strategy for growing tissue is to "print" it with a three-dimensional printer. You can even view a presentation of this on You-Tube, where Dr. Anthony Atala, director of the Wake Forest Institute of Regenerative Medicine, explains the process.[24] First, a scanner is used to capture a three-dimensional image of the kidney that needs to be replaced. A tissue sample about half the size of a postage stamp is used to start the computerized process, after which the printer then creates the replacement kidney layer by layer—similar to printing sheet after sheet of paper with a normal printer—developing a three-dimensional organ over a period of about seven hours. On the same day as his presentation, Dr. Atala used a 3D printer on stage to create a nonfunctioning kidney to illustrate the concept, receiving a rousing round of applause when he showed it to the audience.

Once this technology reaches the application stage, bioengineered kidneys will have a significant impact on transplantations because about 90 percent of people awaiting a transplant are waiting for kidneys. Today, organ transplant patients must stay on a regimen of strong immune system suppressants for the rest of their lives to ensure that their bodies' natural immune systems do not reject the donor organs. Future medical advances, however, promise to eliminate this shortcoming by using patients' own stem cells to regrow all or part of a damaged organ. Bioengineered kidneys, livers, pancreases, and hearts will become reality in the not too distant future.

### REPAIRING DAMAGED TISSUE

Even though bioengineered organs are almost certain to be technologically possible within the coming decades, they might not be the primary means of medical intervention. They could be rendered obsolete by advances in organ repair. Injections of stem cells into damaged heart tissue might be able to regrow healthy tissue, thus avoiding the necessity for invasive heart surgery that, by itself, creates the risk for other unwanted side effects. Again, this procedure has been demonstrated successfully in laboratory animals, so it's only a matter of time until it can be perfected for humans.

### FORTIFYING THE IMMUNE SYSTEM

One of the novel approaches to strengthening the immune system is to create cells that are resistant to attack by free radicals, which are known to be a major contributing factor in aging and in a multitude of diseases.

Free radicals create oxidation in the body that's a bit like rust created by oxidation on an old car. With vehicles, there are several ways to prevent rust damage. One way would be to remove the oxidative stressor—in this case, moisture—from the environment. Another approach is to remove the rust when it occurs and repaint the damaged area to prevent future rusting. You can also treat the metal with a rust protectant such as paint, wax, or a rust inhibitor especially designed for the purpose of preventing oxidation. In 2006, Oxford scientist Mikhail Shchepinov developed a process that accomplishes this same feat on a biological level, protecting cells against oxidation from free radical attacks. Shchepinov, who subsequently founded Retrotope to pursue this technique, used food compounds fortified with hydrogen isotopes to create cells that are unusually resistant to free radicals. His experiments feeding deuterium to fruit flies increased their lifespan by 30 percent. Similar experiments in mice showed no harmful effects of large doses of isotopes.[25]

An isotope is an atom that contains one or more extra neutrons in its nucleus compared to a normal atom of that substance. A hydrogen atom, for example normally contains one proton and one electron, but if a neutron joins the proton in the hydrogen nucleus, it creates an isotope called deuterium. Although it is rare, deuterium occurs naturally. In fact, one out of every 3,000 molecules of seawater contains a deuterium atom instead of a normal hydrogen atom. From an aging perspective, isotopes are important because the covalent bonds holding the electrons to the molecule are up to 80 times stronger in isotopes than in normal atoms or molecules.[26] Fortifying nutrients and healthy cells by replacing their normal hydrogen or carbon atoms with isotopes could dramatically strengthen their immunity to attacks by free radicals, which have been linked to Parkinson's, Alzheimer's, diabetes, cancer, renal disease, and heart disease.

Free radicals do their damage by scavenging electrons off nutrients and other productive molecules, rendering them useless—worse than useless actually, since the plundered molecules then scavenge throughout the body themselves to steal electrons away from other productive molecules. This disruptive effect is so great that most scientists agree free radicals play a major role in aging.

Carbon hydrogen isotopes occur naturally in human cells, but they're relatively rare. During fetal development, mothers lose a disproportionate amount of carbon isotopes because they are transported into the developing fetus. Isotopes of carbon and hydrogen occur naturally in human cells, but they're relatively rare. The current working hypothesis is that the covalent bonds between the heavier isotopes are stronger and make the molecules more resistant to damage. As we age, however, the body contains fewer of these isotopic bonds, which makes the cells more vulnerable to free radical attack.

In theory, foods, such as cereal and milk, could be isotopically fortified with deuterium or carbon isotopes and placed into our food supply. When consumed, these stable isotope-containing molecules would be used as nutrients by the body and replace hydrogen-based molecules with deuterium-based ones. Since these nutrients are utilized by the cells, the deuterium is then transferred to the cells to make them almost impervious to free radical damage.

Retrotope is currently working to develop isotopically fortified compounds to treat specific diseases, but the potential for isotopically fortified compounds is immense. One can envision a future where fruit products and drinks could be isotopically fortified to provide increased immunity to oxidation damage and aging itself. Interestingly, Retrotope started as an anti-aging company, but it realized that the money was in pharmaceuticals rather than prevention, so the company switched its approach from creating isotopically fortified foods to creating pharmaceuticals for specific diseases.

Other methods of returning the immune system to a more youthful level are also likely to appear in the future. With advanced age, some white blood cells lose their ability to kill attacking viruses. When this happens, these cells are not just impotent—they actively reduce the body's available supply of healthy killer cells that ward off infections and viruses. In the future, it may be possible to revitalize aging immune systems by using patients' own cells to grow a supply of healthy white blood cells, which would then be inserted via a transfusion process that would filter out existing killer cells and replace them with younger, healthy ones.

Scientists at the University of Minnesota led by Dan Kaufman MD, PhD, have found that natural killer cells taken from human embryonic stem cells are effective at killing tumors.[27] Specifically, these stem cell–derived killer cells have wiped out breast, prostate, testicular, and brain cancer cells. In the future, it may be possible to create these super-cancer-fighting cells using iPSCs from our own bodies. It's not too hard to envision a scenario where, upon a diagnosis of cancer, a patient's own blood or bone marrow cells might be used to create these super-killer cells to destroy the cancer before it can metastasize or cause any serious damage. It's also possible that in the future researchers may be able to alter the expression of certain genes to increase the body's response to stress, thereby boosting the immune system. A study of fruit flies found that overexpression of the D-GADD45 gene increased the life span, possibly by increasing resistance to free radicals, environmental pollutants, and heat stress.[28] Alternately, decreased expression of the D-GADD45 gene shortened life span.

## AUTOPHAGY

Another relatively new approach to regeneration is called "autophagy"—an exciting alternative to other anti-aging approaches because it doesn't attempt to reverse the age of cells or repair damaged cells. Instead, it simply breaks down old, aged cells and uses the raw materials to grow healthy, youthful cells. In theory, this could transform the entire body, including face and skin, to a more youthful function and appearance. Autophagy could also be selectively used to break down cancer cells into harmless raw materials or to replace damaged heart cells with youthful cells. Since the life span of most types of human cells are measured in weeks or months, only extremely long-lived brain and bone cells would appear to be excluded from the miracles of autophagy—should its power be harnessed to reverse aging.

Lest the significance of this be lost, let's state it another way. Even if all the other approaches fail, *autophagy promises an entirely different way of curing aging and age-related diseases.* If autophagy can be selectively turned on, a patient dying from heart disease could just replace all the old diseased heart cells by breaking the cells down and growing new, young heart cells. There would be no need for surgery or lengthy rehabilitation. The most stubborn cancers could be eradicated by breaking down the cancer cells and using the raw materials to grow healthy cells. The state of the art in autophagy is still years behind stem cell research, but it is nonetheless an exciting field.

## THE NEW "CHINESE MEDICINE"

While most regenerative medicine advances have originated in the United States and the European Union, future advances are likely to come from China, the same nation that brought us some of the world's oldest medicines. Zhou Qi, chief scientist with the stem cell research project at the Chinese Academy of Sciences, says they have progressed faster in stem cell research than any other nation over the past ten years: "We are now close to the day when we will be able to hail a breakthrough in this important technology. China needs five to ten years to shift from basic research to clinical application and another ten years to realize a large-scale clinical application."[29]

China tends to build things on a grand scale and this project is no exception. Construction of a massive medical research center, dubbed "China Medical City," is taking place in Taizhou, a city on the banks of the Yangtze River about 150 miles northwest of Shanghai. When completed, China Medical City will cover about 25 square kilometers (about 6,000 acres or 10 square miles). That's a little less than half the size of Manhattan or roughly the size of San Francisco from the Golden Gate Bridge east to the Bay Bridge and then south to Golden Gate Park. It will be a hub of international research and will house centers for R&D, international conferences, exhibitions, manufacturing, and related support facilities.[30] Several international hospitals will be integrated into the project to attract patients from around the world. According to a Chinese government brochure, it will also include branches of distinguished international medical universities and prestigious research and development institutions.

It's easy to see why China would make this a national priority. The ratio of workers to seniors is declining faster in China than virtually every other nation. By 2050, China will have a ratio of only two citizens aged 15 to 64 to each senior.[31]

While China is going full speed ahead, the rest of the world, and the United States in particular, seems to be stuck in a morass of red tape and antiquated thinking when it comes to medical research.

## FROM DISCOVERY TO CLINIC

Decades ago, it took many years for biomedical breakthroughs to reach clinical application. This was especially true when more than one technological advance (often from a seemingly unrelated field) was necessary to bring the discovery into widespread clinical application. For example, the discovery that blood came in types (A, B, AB, O) was made in 1901. The first human blood transfusion followed six years later, but widespread application was limited because there was no way to store blood. It was three decades before preservation techniques allowed blood to be stored for several days, which would bring blood transfusions into mainstream medicine.

Figure 7.1 shows a timeline of advances in blood transfusions, organ transplants, and stem cell research. Whereas the lag from discovery to medical application was sometimes decades in the past, that lag has shortened considerably in recent years.

As you can see from the right side of this figure, the pace of discoveries related to stem cells has accelerated over the past decade. Already,

# Translating Biomedical Discoveries into Clinical Practice

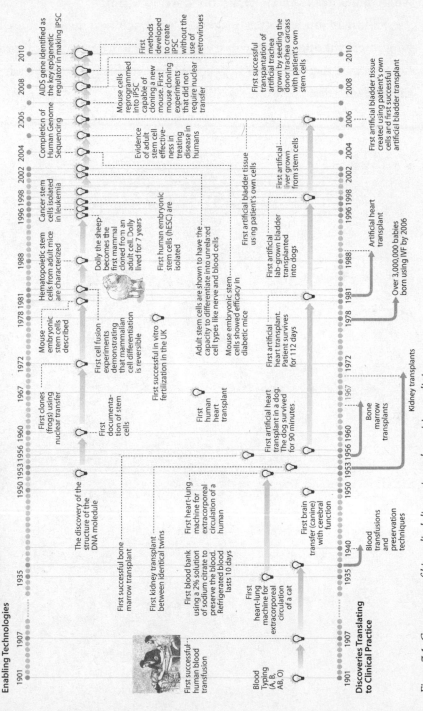

*Figure 7.1. Convergence of biomedical discoveries and transition to clinic.*

some limited clinical applications exist, but we aren't yet at the point where widespread clinical applications are occurring.

Still, it is just a matter of time. Within the next two decades, the floodgates will open to allow into common practice new clinical procedures to cure disease and to slow aging. And it's likely that the leader in these new clinical applications will be China.

# PART THREE

# The Need to Reform Medical Research

While medical knowledge is moving forward rapidly, it is doing so in spite of a morass of bureaucracy and outdated medical research priorities. The way medical research is conducted and funded in the United States needs to be reformed immediately because the expenses of aging and age-related diseases are rising each year. If these expenses are allowed to continue to grow unchecked, they will eventually create a major fiscal crisis that could drag developed nations into a global depression. In fact, the real costs of aging go far beyond medical expenses.

# EIGHT

# The Real Costs of Aging

The unfunded liabilities of old-age pensions and health care for seniors are enormous, but even these costs—potentially leading to the bankruptcy of entire nations—are only the tip of the iceberg. The real costs of aging go far beyond senior entitlement programs, but they are seldom discussed in the same way because almost everyone accepts aging itself as inevitable—just another aspect of life we can't control. But what if we could control aging? Suddenly, the costs of aging would become expense variables that could be managed, controlled, reduced, and in some cases eliminated entirely. That would dramatically change the nature of the discussion.

## NON-SENIOR HEALTH-CARE EXPENSES

We've already discussed at length the substantial financial liabilities that Medicare and other similar programs will face in the future, but they are only part of the cost of age-related health-care expenses. Health-care problems resulting from aging—and subsequent age-related expenses—begin to appear much sooner than age 65. Young adults face very few health problems other than the occasional cold, accident, or athletic injury, but that begins to change with middle age. With age and, most alarmingly, obesity—which affects about one-third of the adult population in most developed nations—more health problems occur. Obesity and lack of exercise—another poor health habit that has become increasingly common over the past half century—are strongly correlated to premature aging and the rise of premature age-related diseases, such as diabetes, arthritis, heart problems, and cancer. These age-related diseases are now appearing with increasing regularity in younger adults, especially among the obese and sedentary population.

By middle age, age-related health problems are already creating significant financial costs for businesses, families, and individuals. Businesses must pay higher health insurance premiums for their employees due to the higher prevalence of illness. Employers also suffer a greater loss of productivity due to absenteeism or the inability of the worker to remain fully productive due to pain management drugs or the debilitating nature of the disease. Governments that provide universal health care, which include almost all European nations, must fund these age-related expenses by increasing taxes or going deeper into debt. Since health care costs are enormous and represent an ever-increasing percentage of each nation's annual GDP, these expenses can be a major limiting factor in future economic growth and overall population well-being.

## OTHER GOVERNMENT COSTS

The costs of aging are not limited to health-care expenses for seniors. Long-term care expenditures are also likely to be significantly higher in the future as more seniors survive previously fatal diseases but require nursing home and rehabilitative care as they live to advanced stages of their disease. Since only a relatively small percentage of those seniors will have enough financial resources to pay for an extended nursing home stay, the burden will fall to the national government, or in the case of the United States, a shared burden between the federal government and states—64 percent paid by the federal government and 36 percent paid by the states. The actual formula varies from state to state, with higher-income states shouldering a slightly higher percentage of the burden, but even so, the financial cost of long-term care has become an increasing concern for states that are already struggling to balance their budgets.

Unlike the federal government, which can print more money when it needs it, states must eventually balance their budgets. Although they can gain a little breathing room by issuing bonds, these bonds are only as valuable as the state's ability to generate revenue in the form of taxes to cover the interest due on those bonds. It's possible that fiscal problems for the United States will begin not as a crisis for the national debt, but at the state level as states begin to default on debts due to their inability to pay off the mounting cost of state employee pensions and the state's share of Medicaid expenses.

The costs of aging are further exacerbated by the costs of public senior housing, senior transportation, an expanded bureaucracy to handle more seniors, and the cost of interest payments on bonds issued to cover national deficits. The United States, for example, pays about $140 billion each year in bond interest payments to its Social Security and Medicare trust funds. When governments spend more money on age-related services, those funds are not available to support other programs. In business, this is referred to as an opportunity cost. Thus, that $140 billion per year spent on bonds held by the Social Security and Medicare trust funds represents a lost opportunity to spend those funds on school nutrition programs, primary- and college-level education, universal health care, and medical or alternative energy research.

And the list doesn't stop there. Money spent on age-related services is money that is no longer available for areas that could potentially improve productivity, such as small-business loans, grants to study advanced manufacturing techniques, or lifelong learning programs that would keep older employees competitive in a rapidly changing workplace. Simply put, aging causes a nation to lose its competitive edge.

## THE COST TO FAMILIES

Families also experience significant financial and emotional costs as a result of aging. Out-of-pocket health-care costs for seniors are outpacing inflation. According to the Employee Benefits Research Institute (EBRI), a retiring couple would need to accumulate $210,000 by age 65 in order to pay their out-of-pocket health-care expenses over their remaining life expectancy.[1] This is in addition to savings needed to maintain their standard of living after retirement. Even more alarming, that figure does *not* include potential out-of-pocket long-term care expenses, which could conceivably top $1 million for a ten-year nursing home stay in some parts of the country.

These financial costs to families are relatively obvious, but there are other less apparent, yet substantial, aging costs. Because they fear losing their health-care insurance, many workers are unable to pursue another job, even if it might pay more or be far more fulfilling. Health problems can make it almost impossible for individuals to take the risk to start their own entrepreneurial business because they would lose health-care insurance for their families. Some families are unable to move to other locations because their current health-care program is only available locally or regionally.

These intangible expenses and similar lost opportunity costs are too numerous to list. Together, they place a growing and unsustainable economic burden on society.

## COSTS TO BUSINESSES

Business costs represent the greatest single area of indirect costs of aging. As workers age and retire, there is a loss of collective corporate memory. Knowledge plus experience—whether a good experience or a bad experience—leads to wisdom, and when senior workers retire, this collective wisdom is lost. Continuity is also sacrificed along with this loss. Depending on the occupation, human resource professionals estimate it costs up to $50,000 to train a new employee. While new employees are getting up to speed, productivity is lost due to their relative lack of experience. The loss of leadership should also not be overlooked. Obviously, given sufficient time and training, younger workers can eventually be as good as, or better than, current workers. The point is, during the time it takes to gain those skills, mistakes are made and productivity is lost.

Of course, aging also creates productivity problems even when workers don't retire. Millions of workers today simultaneously care for aging parents as well as children. The loss of productivity attributed to the latter is called "presenteeism," or a form of mental absenteeism while present, because the worker is distracted by errands or worry over the care of an ailing parent or spouse. Presenteeism can also be the result of stress and less-than-optimal health, both of which can be aggravated by the strains of being a caregiver. According to the Harvard Business Review, presenteeism results in $150 billion of lost productivity each year in the United States alone.[2]

## THE REAL BENEFITS OF ANTIAGING RESEARCH

If aging could be prevented or just delayed a few decades and compressed into a much shorter period of time, all these costs of aging could be ameliorated. Presenteeism would decline; universal health care would be less expensive; Medicare would only need to treat the occasional injury or accident; and businesses would be able to retain their best and brightest employees instead of losing them to old age. Hundreds of billions of dollars could be redirected from health-care expenditures to other quality-of-life issues that would improve the general well-being and productivity of all citizens.

A cure for aging and age-related diseases would also free hundreds of billions of dollars that could instead go to research into providing more nourishing food and clean water to starving nations; solving problems like climate change, pollution, and overfishing; or be used to educate the world's youth. The value of this last benefit should not be summarily dismissed because studies have shown a strong correlation between economic prosperity and the overall educational level of the population.

Although science has made significant headway toward increasing productive years and compressing morbidity (the number of years of declining health before death), this is true mostly for individuals who proactively pursue good health. Millions of adults will still experience premature aging due to obesity, poor lifestyle habits, or just poor genes. The sooner medical science recognizes aging as a disease and takes aggressive steps to prevent or cure it, the greater becomes the probability that we can survive and prosper without the threat of an apocalyptic collapse looming on the horizon due to the costs of senior entitlement programs.

## SOCIETY SECURITY

The Social Security program began running an annual deficit for the first time in 2010, seven years sooner than official forecasts. This has been blamed on higher-than-anticipated unemployment rates that resulted in fewer employed workers paying into the system, but that was only part of the problem. High unemployment caused a higher percentage of older workers to file for early Social Security benefits. As a result, it is highly unlikely that the Social Security program will ever have a surplus again. Instead, deficits will continue to grow as millions of baby boomers become eligible for Social Security benefits each year. The official position, as stated in the 2012 Social Security Trustees Report, is that the Social Security trust fund can cover these deficits until 2033, but the reality is quite different.[3]

The Social Security trust fund—although it exists on paper—is a troubling mix of IOUs and accounting sleight of hand that hides the true depth of the crisis. Officially, the Social Security trust fund works by investing its annual surpluses into securities, but in reality, the treasury doesn't actually leave that money in the trust fund. Instead, the government spends the money on other federal programs, such as defense, education, and welfare. The effect is similar to replacing money in a piggy bank with a handwritten IOU. When funds are needed, there will be no real money in the account. Similarly, when the Social Security trust fund attempts to redeem its treasury securities, the U.S. Treasury will be forced to issue new treasury securities to replace them. When that happens, those new bonds will increase the public debt of the United States as if the Social Security trust fund never existed.

**Summary of 2010 Trust Fund Financial Operations**
[In billions]

|  | OASI | DI | OASDI |
|---|---|---|---|
| Assets at the end of 2009 . . . . . . . . . . . . . . . . . . . . . . . | $2,336.8 | $203.5 | $2,540.3 |
| Total income in 2010 . . . . . . . . . . . . . . . . . . . . . . . . . | 677.1 | 104.0 | 781.1 |
| Net payroll tax contributions . . . . . . . . . . . . . . . . . . . | 544.8 | 92.5 | 637.3 |
| Reimbursements from General Fund of the Treasury . . | 2.0 | .4 | 2.4 |
| Taxation of benefits . . . . . . . . . . . . . . . . . . . . . . . . . | 22.1 | 1.9 | 23.9 |
| Interest . . . . . . . . . . . . . . . . . . . . . . . . . . . . . . . . . | 108.2 | 9.3 | 117.5 |
| Total expenditures in 2010 . . . . . . . . . . . . . . . . . . . . | 584.9 | 127.7 | 712.5 |
| Benefit payments . . . . . . . . . . . . . . . . . . . . . . . . . . | 577.4 | 124.2 | 701.6 |
| Railroad Retirement financial interchange . . . . . . . . . . | 3.9 | .5 | 4.4 |
| Administrative expenses . . . . . . . . . . . . . . . . . . . . . . | 3.5 | 3.0 | 6.5 |
| Net increase in assets in 2010 . . . . . . . . . . . . . . . . . . | 92.2 | -23.6 | 68.6 |
| Assets at the end of 2010 . . . . . . . . . . . . . . . . . . . . . . | 2,429.0 | 179.9 | 2,609.0 |

Note: Totals do not necessarily equal the sums of rounded components.

*Figure 8.1. Sample SSA Trustee Report: Summary of 2010 Trust Fund Operations.*

In 2011, for example, the Social Security trust fund received about $114.4 billion in interest from the treasury bonds it holds, but that is extremely misleading. This money essentially amounts to another IOU, one to pay the interest on the previous IOUs. The treasury must raise that $114.4 billion by issuing new treasury bonds. When it issues new treasury bonds, those bonds further increase the national debt.

Thus, the financial problems of the Social Security program are much greater than reported. The 2012 Social Security Trustees Report shows a surplus of $69 billion, but Table II.B1 of that report shows $102.7 billion in "income" from "Reimbursements from General Fund of the Treasury." That's general revenue tax dollars Congress put into the trust fund to cover its shortfall. The same table also shows $114.4 billion in bond interest as revenue, but in reality, new bonds had to be issued to pay that bond interest. In a real business, that $114.4 billion on bond interest and $102.7 billion in general revenue would not be counted as income. Therefore, when we remove this curtain of creative accounting, the Social Security trust fund did not operate with a surplus of $69 billion, as reported in the 2012 report. In reality, it ran at a deficit of $148.1 billion.[4]

## MEDICARE

As bad as the Social Security numbers look, Medicare's are even worse. The 2011 Medicare Trustees Report projected that the Medicare Hospital Insurance (HI) trust fund would remain solvent until 2029. The 2012 report accelerated that crisis date to 2024, but these reports only refer to the HI portion of Medicare, *not the entire Medicare program.* In reality, the overall Medicare program is *already running a massive deficit of over one-quarter of $1 trillion per year.* Table II.B1 of the 2012 Medicare Trustees Report (coincidentally the same table number as the Social Security report) claims a Net Change in Assets of negative-$19.2 billion, but a closer look at the items under "Income" shows a line item for "General Revenue" of $223.3 billion.[5] General revenue is money from federal income taxes—that is, money obtained from taxpayers.

By law, Congress must anticipate any shortfall in Medicare in the upcoming year and automatically transfer sufficient funds from general revenue to meet that Medicare shortfall. That is why general revenue funds are shown under income, even though it would seem to be more appropriate to just show an operating loss for Medicare.

General revenue isn't really a source of income for the trust fund; it's really general tax dollars. As we discussed earlier, interest income from bonds held by the trust fund isn't really income either, so it should also be removed from the income category. After correcting for this, the operational loss for Medicare was not $19.2 billion, but $19.2 billion plus $223.3 billion, or $241.5 billion. Medicare also received $15.2 billion in bond interest that year. Since new bonds had to be issued to make that $15.2 billion interest payment, Medicare's true deficit that year was actually $15.2 billion plus $241.5 billion, or $256.7 billion. Thus, as politicians and official trustee reports assure the American public that Medicare will be okay until 2024 and Social Security will last until 2033, *these two programs are currently running an annual shortfall of over $400 billion.* Worse, that figure is increasing each year.

In 2011, the U.S. national debt reached $14.4 trillion, topping 100 percent of GDP, a level surpassed briefly only once before, during World War II. By 2012, it had grown to $16 trillion. Social Security and Medicare are a major part of this overspending—adding $400 billion each year. Worse, the combined deficit from these programs will grow even faster as more baby boomers become eligible for these programs.

How long can the United States sustain this deficit spending without catastrophic consequences?

## OTHER NATIONS, SIMILAR PROBLEMS

Today, Social Security is the larger item in the national budget—$800 billion in 2010 in comparison to $600 billion for the next largest budget item: Medicare/Medicaid. Other nations face similar problems. Japan, for example, spends 7.5 percent of GDP on old-age pensions compared to 5.3 percent by the United States. Spending on old-age pensions in Italy and France is even higher at 11.6 percent and 10.7 percent respectively. In terms of spending on old-age pensions as a percent of GDP, the United States is in a better position than about half the nations in the European Union.[6]

## SUSTAINABILITY OF SENIOR PROGRAMS

Many experts have expressed justifiable concerns over the long-term viability of government-sponsored old-age pensions and health care. Most of the developed nations in the world are seeing substantial increases in the numbers of their elderly, while health-care expenses continue to exceed the rate of inflation. An even more ominous trend is the declining ratio of workers to seniors, resulting in fewer workers contributing into these programs while the number of seniors receiving benefits is projected to double over the next few decades.

Many countries are already running annual deficits each year and have accumulated sizable national debt, to the point that just paying the interest on this debt is a significant part of their annual budgets. As we saw earlier in the case of the United States, the portion of this national debt due to senior programs is substantial and getting higher each year. The question becomes whether or not nations can continue to run these deficits indefinitely. If not, what are the options for managing these deficits?

## MANAGING EXPENSES

As with any budget, there are two primary variables—expenses and income—that can be adjusted. Unfortunately, dramatically reducing expenses would seem to be out of the question. Life expectancy continues to increase. The senior population is growing by millions each year. Medicare expenses have increased over 10 percent per year for the past year, more than triple the annual inflation rate for the same period. Worse, the obesity epidemic is poised to further increase Medicare expenditures.

According to a study of the relationship between one's weight in middle age and Medicare costs later in life, Martha L. Daviglus, MD, PhD, of the Feinberg School of Medicine at Northwestern University, found that obese seniors who were also obese in middle age had 50 percent higher Medicare expenses than their normal-weight peers.[7] This is an ominous harbinger of future problems for Medicare because while only 23 percent of today's seniors are obese, the percentage is a much higher 34 percent in the working-age population and increasing each year.[8] Among baby boomers, obesity is even more prevalent—topping 40 percent and increasing at roughly 1 percent per year.[9]

## MANAGING INCOME

As we said earlier, there are two major variables with any budget: income and expenses. Since it appears expenses will inevitably rise in the future, what can be done to increase the income available to fund these programs? One obvious approach would be to increase FICA (Federal Insurance Contributions Act) withholding. An increase of $2,000 per worker would do it temporarily, but that would effectively double the net tax rates of low-income workers. Even then, the solution would only be temporary because withholding would have to be increased by about $200 each year per worker just to keep up with Medicare inflation and the rising number of seniors.

Another option would be to increase Medicare premiums, about $1,300 per year in 2013, or to increase premiums for seniors with higher incomes.[10] Again, this would barely make a dent in the Medicare shortfall because Medicare expenses currently average about $12,000 per beneficiary per year.[11] Substantial increases in Medicare premiums would create even more problems because many seniors are barely able to survive on Social Security alone.

## PUTTING IT IN PERSPECTIVE

America's national debt has become so large that it is hard to put into terms that the average person can comprehend. Perhaps it's easier to think about in terms of federal income taxes. Forget for the moment that the national debt of the United States is over $16 trillion. For this illustration, let's also forget that wars in Iraq, Afghanistan, and other hostile locales have cost over $1.5 trillion. Let's simply look at the expenditures of Social Security and Medicare. How much would individual federal income taxes need to increase to balance the annual budgets of these two programs?

In 2009, the combined deficit of these two programs plus the interest on the bonds held by their respective trust funds was about $208 billion. In order to balance the budget that year for those two programs alone, tax revenues per worker would have had to increase by about $1,600. The following year, these two programs incurred an even larger annual deficit of about $420 billion. To balance the budget that year, the taxes per worker would have had to increase by an additional $1,700. The average worker would then be paying $3,300 more per year in taxes than in 2008. In 2011, baby boomers continued to sign up for early Social Security benefits in record numbers, and nearly 4 million baby boomers joined the ranks of Medicare recipients. Of those 4 million, 67 percent were overweight, 40 percent were obese, and 16 percent were severely obese, further increasing the likelihood of even more Medicare expenditures in the future. To offset the additional deficit incurred by Medicare and Social Security in 2011, taxation of the typical worker would have to increase another $330.

Before moving to 2012 and beyond, let's summarize the situation. To balance the budgets of Social Security and Medicare from 2009 to 2011, the average taxation would have to have been increased by over $3,600 per worker. That's only for Social Security and Medicare deficits, not the overall deficit. That's only for three years and only the average tax increase. Many workers would have to pay substantially more in order to offset the lower tax rates for lower-income workers. Meanwhile, overspending in other areas of the budget—bank bailouts, mortgage bailouts, military spending, stimulus programs, and so on—added another $4 trillion to the national debt during those same three years. To balance the budget for all that spending, the average worker would have to pay an additional $31,000 in taxes over those three years.[12]

Bear in mind, all these steps only balance the budget from 2009 through 2011. This exercise doesn't address any of the national debts accumulated prior to 2009. Plus, it only gets the United States through three years of deficit spending.

Some skeptics will argue that this scenario is overly pessimistic. They'll say the economy could recover, thus adding more workers. Medicare and Social Security could be reformed, reducing future expenses to more manageable levels. But that's the whole point—of course these programs must be reformed, because they are unsustainable as they exist today. It's only a matter of time until continued deficits create a financial crisis that would be catastrophic for our way of life, not just in the United States but in other nations around the world.

So far, we have only discussed Social Security and Medicare. There are other risks we haven't even touched on that, when considered, paint an even bleaker future for the United States. From 2009 to 2011, the cost of gasoline doubled, adding about $1,500 to the typical family's an-nual transportation expenses. In the future, each additional $1 increase in gasoline prices adds about $1,000 to the average household's expenses, which will drive many families even deeper into debt.

Increasing national debt also weakens the dollar on the international market, making imported goods, such as oil, rise in price even more. Increasing debt also makes it harder to sell our bonds because buyers fear that, at some point, a crisis might be reached that would cause the United States to default on its bond payments. Fear of the dollar declin-ing in value inevitably leads to fewer bond buyers. To attract buyers, the bond market must raise the interest rates paid on U.S. treasury bonds until these higher interest rates attract buyers back into the bond market. Rising interest rates on bonds have a ripple effect throughout the United States, the rest of the world, and ultimately, on the typical household. While you may not personally own bonds, the interest you pay on a car loan or home mortgage will go up as interest rates rise, because the cost of commodities—metals, produce, and land—will increase accordingly.

## RAISE TAXES

Some experts argue that the solution is to increase taxes on those house-holds with higher incomes. Other experts assert that increased tax rates stifle growth, slow the economy, and result in much less tax revenue than anticipated. It is very difficult to find an objective financial analysis of how much revenue could be generated at various rates of taxation. Since this data is critical to any informed decision on taxation, I collaborated with a finance professional to create the following analysis.

The United States currently spends about $310 billion per month, while its revenue from all sources is about $175 billion per month. That leaves a monthly shortfall of $135 billion.

The top 1 percent of wage earners (those with incomes greater than $380,000) pay a net effective tax rate of 23 percent after all deduc-tions. The next 9 percent of wage earners (those with incomes between

$113,000 and $380,000) pay a net effective tax rate of 18 percent.[13] "Net effective rate" is the actual rate of taxation after all deductions are taken.

If tax loopholes were eliminated and tax brackets raised enough to change the net effective tax rate of the top 1 percent of wage earners from 23 percent to 33 percent (a 44 percent increase in taxation), it would generate an additional $170 billion annually in revenue, or $14 billion per month.

If the same steps were taken for the next 9 percent of wage earners ($113,000 to $380,000) to increase their effective net tax rate from 18 percent to 23 percent (a 28 percent increase in taxation), it would generate an additional $105 billion in revenue, or about $9 billion per month.

Politically, it would be difficult to pass tax increases of this magnitude, but even if it could be done, the combined increase would only result in $23 billion more revenue per month. Meanwhile, the current shortfall in the United States is $135 billion per month. Since these levels of taxation don't provide a meaningful dent in the deficit, could the government double the net taxation rate for everyone making over $113,000? At that rate, someone earning $113,000 per year would be paying over $40,000 a year in taxes! This is such an extreme increase in taxation that it is unrealistic, but if it were done, it would result in $395 billion per year more revenue from the top 1 percent and $375 billion more from the next 9 percent of wage earners. The total would be $770 billion, or $64 billion per month. This would reduce deficit spending by less than half—from $135 billion to $71 billion per month.

The key point here is that *doubling the net effective tax rates on all workers earning more than $113,000 per year would cover less than half of current deficit spending.*

## A COMING CRISIS?

The United States is currently running a deficit of about $1.6 trillion each year. National debt stands at over $16 trillion and is on pace to top $30 trillion in less than ten years. Interest payments alone on that debt will be nearly $1 trillion. Already, the two largest bond buyers—China and Japan—have begun to reduce their annual purchases of U.S. Treasury bonds. This is significant because if there aren't enough buyers at each monthly treasury bond auction, the U.S. Treasury has to raise treasury bond interest rates to find buyers. Higher interest rates means the United States would go even deeper in debt and be forced to sell even more bonds, which in turn would drive interest rates even higher. This hasn't happened yet, but in other nations the resulting spiral of printing more money and issuing more debt has created hyperinflation, very high unemployment, civil unrest, and has even resulted in overthrown governments, such as in the cases of Bolivia and Argentina.[14] In spite of its size, the United States is not immune to these risks. With the interconnectedness of economies today, if the United States plunged into a true depression, it could engulf the civilized world in a financial crisis far greater than the Great Depression of the 1930s. It could become the worst social disaster since the Dark Ages, when millions starved or froze to death from lack of heat. Today's concerns—clean energy, clean water, reduced dependency on oil, climate change, world hunger, environmental pollution, and more—would be rendered irrelevant *because no funds would exist to support these programs.* No funds would exist to support *any* government programs. Some nations would spiral into chaos and civil war. At least during the Great Depression, people had extended families and neighbors who could help out. In today's urban and suburban isolation, most developed nations have lost the sense of community that brings people together during times of real crisis. In the Great Depression, most rural families were able to grow their own crops and raise farm animals for food and cut wood for cooking and heating. Today, far fewer families have this ability.

The solution, as I have previously explained, is to extend the health of seniors to the point that 65 is no longer "old." If seniors were healthy and working, spending on senior welfare programs could be far less than it is today.

# NINE

# Changing the Priorities
# of Medical Research

I n 1971, President Richard Nixon declared war on cancer, passing the
National Cancer Act and proclaiming that we would find a cure for the
disease within five years. It was a bold plan, but it was an era of bold res-
olutions. President Lyndon Johnson's Great Society had recently expanded
welfare and created Medicare and Medicaid. Barely a decade earlier, President John F. Kennedy made an equally bold proclamation to put a man
on the moon. It took only eight years to reach that goal, but 40 years after
the passage of the National Cancer Act, the results are, at best, mixed. Advances have been made in early detection, but very few types of cancer can
be cured outright. In spite of billions spent on research, the yearly number
of new cancer cases per 100,000 people increased in the United States from
400 to 468 from 1975 to 2009.[1] The age-adjusted cancer death rate has
declined less than 10 percent since 1975, with a significant portion of that
decline coming from fewer smokers.[2] It would seem that, by these measures
at least, Nixon's war on cancer has failed. In a large part, the responsibility
for that failure lies with the National Institutes of Health (NIH).

## THE NATIONAL INSTITUTE OF HEALTH

The NIH is the major research funding arm of the United States, spending nearly $31 billion on medical research in 2011. But over the past 20

years, less than 2 percent of its expenditures have gone into research projects in regenerative medicine—one of the most promising areas of aging research.[3] According to the NIH's website, its mission is "to seek fundamental knowledge . . . to enhance health, lengthen life, and reduce the burdens of illness and disability." The goals of the agency are to "foster fundamental creative discoveries, innovative research strategies, and their applications as a basis for alternately protecting and improving health; to develop and maintain and renew scientific human and physical resources that will ensure the nation's capability to prevent disease . . . by conducting and supporting research in the causes, diagnosis, prevention and cure of human diseases; the processes of human growth and development; the biological effects of environmental contaminants; the understanding of mental addictive and physical disorders."[4]

Those objectives are broad, altruistic, and humanitarian. Essentially, it's what one would expect from a national institute, but times have changed. As the United States and other developed nations are forced to take a hard look at spending cuts to bring their mounting deficits under control, every national agency will face potential budget reductions, and the NIH will be no exception. The United States can no longer afford to spend lavishly on medical research for the sake of pure research. With limited funding, it becomes essential to place a higher priority on research that can potentially have a meaningful impact on overall health or health-care expenses.

The NIH comprises 27 separate research departments, all listing vague, lofty-sounding goals on their websites. These goals and objectives are so broad that one might think that if the word "health" can somehow be worked into a research grant application, it stands a reasonable shot at getting funding. That's a bit of an exaggeration, of course, but it's also apparent from the stated objectives that research on aging, cures, or dramatic reductions in Medicare costs do *not* receive higher priority than any other type of research.

As a result, researchers sometimes look to apply for a grant based on novelty or current fads rather than projects that could provide meaningful financial or medical benefit. One such research project was a grant to study the effects of barbershops on hypertension among African-American males. This is not a joke—it was a real study.[5]

Another study—details altered to disguise the identity of researchers involved—received funding to track the long-term effects of stress on elderly women who have had hip surgery.[6] Several hundred older women were identified and monitored after hip surgery. They completed surveys, annual checkups, and were monitored for other variables that could skew the research—subsequent accidents, complications, the rise of new diseases, death of a spouse, and so on—anything that might impact the study of stress. At the end of ten years and at a cost of millions of dollars, the study draws these conclusions: hip surgery leads to stress; stress leads to other illnesses; these illnesses in turn increase the risks of morbidity and mortality. Research projects like this pop up every so often in the news, leading inevitably for people to remark, "They needed a study to find out *that*?"

Another problem is that roughly 70 percent of NIH funding goes to long-term multiyear research projects. These projects are often social- or behavior-oriented and can be very expensive to conduct. Decades ago, it was possible to set up such multiyear studies without fear that the underlying variables might change rapidly and thus invalidate the entire study. Today, however, the pace of technology has accelerated to the point that it is possible for a technological advance to render a study meaningless before it concludes. Unfortunately, once these long-term projects are started, annual funding becomes almost automatic.

To illustrate this, let's look at a hypothetical study tracking thousands of seniors for ten years with the purpose of evaluating the impact of social interaction on their overall health. To do this, researchers might measure the size of their immediate circle of friends, the frequency of visits from relatives and visits to see others, the number of letters sent and received, and number of phone calls made. Today's new social media, however, makes this hypothetical study obsolete. Instead of writing letters to grandchildren, seniors can connect via Facebook, and can even

see their families via video calls on Skype, making phone calls unnecessary. Overall, the quality of interaction with distant relatives has been
greatly enhanced by the new social media, but phone calls, letters, and
visits have most likely decreased. The metrics of this study have changed
so dramatically that the outcome becomes irrelevant without changing
the study's parameters. Yet in spite of this obvious outcome, dozens of
similar real-life research studies will continue to be funded automatically
each year until completion, draining millions of valuable research dollars
away from research on cures, prevention, and regenerative medicine. The
solution is to reevaluate long-term studies periodically to ensure the purpose of the study has not become compromised or rendered unreliable by
changing events and technology.

Granted, most NIH research projects have a worthwhile purpose,
but that doesn't mean that they should be funded or, in the case of existing long-term projects, continue to be funded. In view of the looming
health crisis caused by aging and age-related diseases, the NIH should
establish a new set of standards that place a higher priority on goals for
the greater good to humanity. Top priority should be given to research
projects that show promise of increasing the number of healthy work
years, curing major diseases, slowing the aging process, or significantly
reducing Medicare costs.

Specific milestones should be established and updated every three to
five years to ensure continued relevancy. For instance, a goal might be to
extend healthy work span by 10 years over the next 20 years. (This might
sound far-fetched, but China has already established a similar goal. As
you may recall from chapter 1, China recently announced plans to invest $308.5 billion in biotechnology over the next five years with specific
goals of generating 1 million jobs, extending life expectancy by one year,
and reducing childhood mortality by 88 percent.) Once new criteria are
established, research projects that fall outside of these parameters would
be eligible for funding only after higher-priority projects were considered.
Alternately, a department might allocate a small percent of its budget for
other worthwhile projects that did not meet these top criteria.

If priority were given to projects that showed promise of curing major diseases, slowing the aging process, or significantly reducing Medicare, one-third to one-half of existing projects would fail to meet these criteria. A realignment of priorities could free up over $15 billion in the NIH budget to allocate for immediate research in cures, prevention, and re-generative medicine. And the projects that fall outside of this spectrum could still seek funding from private sources or from public nonprofit organizations, such as the American Cancer Society.

This approach would necessarily eliminate funding for many worth-while projects, but with limited research funding, it makes more sense to fund projects that have the greatest potential value to humanity. Re-searchers would adjust their grant proposals accordingly to place more focus on the big picture. If research into a rare form of cancer might possibly provide proof of concept of a new approach to treating common forms of cancer, then researchers should state that in their grant proposal, which could elevate their project to a higher priority category.

Here is a real-world example picked at random to illustrate this dis-cussion. Askin tumor is a small-cell cancer of the chest wall. It is very rare—only about 100 cases are identified in the United States each year. A search through AgingPortfolio.org found an ongoing series of research projects on this rare disease totaling $1.2 million in funding over the past four years. Could these research dollars be better spent, perhaps on a proj-ect to discover an overall cure for cancer? Difficult questions like this one will have to be asked for hundreds of research projects if NIH funding is reduced in the future.

## ENTRENCHED VIEWS ON AGING

Another problem has little to do with the system and more to do with the scientists and researchers within it. When talking to professionals in gerontology, aging, and related fields, it's not uncommon to hear the as-sertion that aging is a beautiful process and that it would be unethical, even unnatural, to interfere with it. I've heard such comments myself at conferences from people who are held in high esteem within their respective fields. These comments would be laughable if the stakes weren't so high. Aging is a natural process? Cancer is a natural process, too—cancerous cells are created every day in our bodies. Would it be unethical and unnatural to interfere with cancer? Of course not. So why should we feel any different about aging?

## LACK OF CLEAR DEFINITION OF AGING RESEARCH

While working on the International Aging Research Portfolio, I noticed another major impediment to aging research: because aging is a multidisciplinary process, it's difficult to define whether or not a given research project is actually related to aging. Sometimes scientists themselves are confused about their own contributions to aging research.

At a recent event hosted by a close friend of mine, Dr. Charles Cantor, former director of the Human Genome Project, I was introduced to a brilliant young scientist from Scripps, Dr. Kristin Baldwin. In 2004, Dr. Baldwin became the first scientist to clone a living mouse using mesenchymal stem cells. She then repeated this remarkable process using induced pluripotent stem cells in 2008. Her revolutionary experiment has immense implications for aging research because it demonstrates proof of concept that mesenchymal cells can be engineered to grow into an organ or even an entire living organism. If it can be done with a mouse, it's possible that it can eventually be done with humans.

I complimented Dr. Baldwin on her accomplishments by remarking that she was one of the most prominent scientists in aging research, but she quickly corrected me, pointing out that she does not work in that specific field: "I'm a cell biologist, neurobiologist, and molecular biologist," she replied, "not an aging scientist." She added that she had no scientific interest in aging. I was shocked because her research had such obvious implications for aging and regenerative medicine. Later, during my travels to China, I met with several scientists who had replicated and improved upon Dr. Baldwin's protocols for making iPSCs and turning them into living mice. There was no confusion among this group; these scientists clearly identified themselves as working in regenerative medicine and aging.

These two extreme positions illustrate a major problem for aging research funding. Because research is going on in so many different fields, it is difficult to tell what constitutes aging research and what does not. Of course, if aging research was given higher priority for funding, this problem would disappear because scientists would refocus their research to emphasize its anti-aging benefits. Just by slightly changing the focal point of research projects could result in significant regenerative medicine breakthroughs over time as thousands of scientists start to actively look for anti-aging implications in their research projects.

## THE PHARMACEUTICAL INDUSTRY

One of the most formidable obstacles in regenerative medicine research is the pharmaceutical industry. In the United States, this industry makes substantial contributions to both political parties, thus ensuring their political influence no matter which party is elected. In the 2008 presidential election, pharmaceutical companies donated nearly $27 million to presidential candidates—more contributions than any other industry.[7]

In itself, that's not a bad thing—in fact, making political donations is a financially prudent step for companies with literally billions of dollars in profits hanging in the balance on the somewhat arbitrary whims of politicians. The mission of any corporation is to increase profits for its shareholders. In this respect, pharmaceutical companies are no different from any other corporation, except for the fact that they have accomplished their mission exceedingly well. Over time, the financial and political influence of the pharmaceutical industry has become so great that it has distorted free-market forces and the fundamental structure of medical research, not just in the United States, but throughout the world.

## UNPRODUCTIVE DRUG RESEARCH

Each year, billions of dollars are spent on drug research to patent products that increase profits for pharmaceutical companies, yet provide little or no added benefit in managing a particular disease. Ironically, these research dollars are not just wasted; they often add to the cost of health care by increasing the cost of drugs when equally effective generic drugs are available. Dozens of inefficiencies exist within the drug research pipeline, but we'll focus on just three major problems that collectively cost Medicare and similar senior health schemes around the world billions of dollars each year.

### *Managing Symptoms, Not Researching Cures:*

The overwhelming body of drug research focuses on finding ways to manage or control the symptoms of the disease rather than cure the disease or prevent it from occurring. Managing disease provides ongoing profits for pharmaceutical companies. Researching cures would be the equivalent of killing the goose that lays the golden eggs.

### *"Me Too" Drugs:*

A "me too" drug is a new drug that provides a similar function to another drug scheduled to go off patent in the near future. When a drug goes off patent, generics can be created that dramatically reduces profits for the original drug. To avoid this, pharmaceutical companies develop new, but almost identical drugs—perhaps with nothing more than a molecule change here or there. They can then obtain a new patent with an exclusive multiyear protection to produce their new drug. Then, they mount a massive marketing campaign to convince the public that their new drug is better than their old drug.

Surprisingly, the FDA places no restrictions on "me too" drugs. In fact, a new drug does not even have to be significantly better than existing drugs to receive FDA approval. It must only be statistically better than a placebo. Obviously, changing a molecule in an already existing drug requires far less time and expense than researching a brand-new drug, so significant amounts of precious research dollars get funneled into wasteful research.

### *Evergreening:*

Another way to expand multiyear patent protection on an old drug is to change the delivery method—from oral to topical, for example. These new drugs are virtually identical to the previous drug except for the delivery method. As a result, the pharmaceutical company gains multiyear patent protection on the "new" drug, which is really the old drug, but they can market it heavily to convince physicians and the public that its benefits make it better than the old version. The cost of these new patent-protected drugs gets passed on to consumers, Medicare, and other health insurance plans.

To be sure, even more problems exist with the drug research system. Another shortcoming is the allocation of research dollars toward drugs that are incredibly expensive but provide only marginal benefits. As case in point, two drugs were recently brought to market by well-known pharmaceutical companies to treat late-stage radiation-resistant prostate cancer. These drugs are only marginally more effective than a placebo, extending the remaining lifetime of these terminally ill patients from 11 to 15 months. For those aware of end-stage prostate cancer, those extra four months are not likely to be very pleasant. The cost to research, develop, and bring these drugs to market was tremendous—over $1 billion. Because these are last-recourse drugs, they are typically given to only 10,000 patients per year. Research for similar limited-use drugs is ongoing around the world right now at a cost of billions of dollars each year. Wouldn't it be more prudent to redirect those billions of dollars into research to find a vaccination or cure for all prostate cancers? The answer is self-evident. So why are other drugs with similarly limited benefits receiving billions in research funding each year? The answer is equally obvious—they generate massive profits for pharmaceutical companies.

Given the economic realities facing Medicare, the most compassionate and prudent approach would be to reform research funding so that future dollars are spent on projects that provide substantial health benefits. Those projects would necessarily include finding cures or vaccines for major diseases and slowing the aging process.

These are only a few of the inefficiencies in the current system of drug research. The arrival of less expensive generic drugs is often delayed by a year or more because of legal loopholes, allowing the pharmaceutical company to generate billions of dollars in additional revenue during the extension period before the patent expires. Another common problem is that the same drugs cost *more* in the United States than they do overseas *even when they are made by the same company under the same conditions.* All of these inefficiencies represent opportunities to generate savings that could be better utilized to research cures for major diseases and ways to slow aging.

## INTERNATIONAL EFFORTS

In the current political and economic environment, there are four major centers of medical research: the United States, European Union, China, and Japan. Of these, China is by far the most aggressive in pursuing regenerative medicine, having committed in 2011 to spending $308 billion over five years on biomedicine.[8] Since China provides much less support for projects of questionable value, like the social/behavioral projects discussed previously, their research dollars are likely to go to projects that can significantly improve health and longevity. As a result, China could assume the world's leadership role in regenerative medicine research within the next decade.

Meanwhile, the United States and the European Union are proposing austerity programs that include reductions in many areas of science, including research. To stay competitive with China, the United States and the European Union will immediately need to reform the current structure of medical research. The new focus should be on curing and preventing diseases, extending healthy life span, and reducing health-care costs. All other ongoing multi-year projects should be reevaluated based on these criteria. If they don't meet the criteria, they should be defunded to free up more funds for higher priority goals.

If the societies of developed nations are to survive and prosper in the coming decades, governments must accelerate research into medical fields that promise to slow or even stop the aging process. We need a bold new initiative—an Apollo program on aging and age-related diseases. There are no other realistic solutions to resolve the financial problems of senior health care over the long term. Medicare expenditures would need to be cut by 40 percent just to offset today's deficits.[9] Increasing taxes enough to cover that shortfall isn't realistic either. There's not enough revenue to tax.

# PART FOUR

# The Retirement Culture

Even if medical breakthroughs can soon postpone physical aging, many other obstacles must be addressed if we are to change the current retirement culture. Radical shifts in aging will require radical shifts in thinking—not just about aging, but about all aspects of society. Perhaps the most obvious question is what to do about retirement. With life expectancy increasing rapidly, many of today's 65-year-olds in the United States will receive pensions for 20 years or more. Among European nations with lower retirement ages, the financial burden of their pension schemes will be even greater. As aging populations grow faster than young people join the workforce, senior entitlement programs will become increasingly unsustainable.

In 2012, an article coauthored by myself and other concerned professionals in *Pensions International Journal* analyzed the economic impact of rapidly decreasing mortality in great detail.[1] We found that far higher numbers of seniors reaching advanced age will place a significantly higher financial burden on Social Security and Medicare than official estimates. Something will have to change.

Extending the age of eligibility for old-age pensions and senior health care would alleviate these burdens temporarily, but changing the entrenched retirement culture has proven to be very difficult, as evidenced by protests in France when the government increased retirement age from 60 to 62. Obviously, it is in every nation's best interests to keep seniors healthy and working longer. The longer seniors pay into these programs before starting to withdraw from them, the longer these programs can survive. In turn, this would allow more time for medical advances to further extend the health span (and thus the working years) of seniors. That, in turn, would extend the productive years of older workers even more. Changing the existing retirement culture to accommodate working to advanced age will be difficult, but it may come as a surprise to some readers that the modern concept of retirement is a relatively new construct.

# TEN

# Changing the
# Retirement Culture

The current retirement culture is rapidly becoming obsolete in the new reality of the twenty-first century. For the past 50 years, retirement has been depicted as a lifestyle of leisure and luxury—playing golf in tropical settings, traveling around the world, dining at exotic places, and generally living the good life. Alternately, of course, there has always been a segment of society that will never reach these lofty goals. Retirement instead will be a time of forced inactivity due to lack of funds or financial insecurity, not knowing when the next medical bill could devastate marginal retirement savings.

With the changing financial environment and increasing longevity, these two scenarios no longer reflect the reality for seniors, except for the extremes of the bell curve—the very rich and those who are unfortunate enough, either through life circumstances or poor health, to remain unemployed after age 65. For the vast majority of the population, working past age 65—in some cases, far beyond age 65—will become the new norm in the twenty-first century.

Unfortunately, large sections of the population still cling to the old paradigm, reinforced by marketing campaigns of the financial services industry providing retirement planning services and print media advertising for upscale retirement living communities. "You've earned a good life, or if you haven't yet, we can help you achieve it"—it's the recurring theme of these commercials and it is a powerful cultural force. Some seniors will, in fact, achieve those dreams, so it's not totally out of the question, but the dream of a retirement of leisure beginning at age 65 is simply no longer realistic for the vast majority of workers. A middle-income 65-year-old American couple would need over $1 million in accumulated retirement savings to achieve a secure retirement without significantly reducing their standard of living.[1] The tragedy is that the majority of 65-year-old couples have less than half that in savings when they reach age 65.[2]

Over the last five decades, retirement has become part of the American dream and the established expectation of seniors in virtually every major society in the world. Yet it should be painfully obvious that the retirement status quo will have to change in the near future or society will soon experience the consequences of runaway deficits. History has taught us that once the foundation has been laid by ongoing deficits and massive national debt, a major financial crisis can arise quite suddenly.

With increasing longevity, it will be essential to change the retirement culture to realize the financial savings possible from delaying old-age pensions. Changing the retirement age would also help prevent millions of families from retiring prematurely only to find a few years later that their retirement savings were woefully inadequate. However, bringing about that change will not be easy.

An entire generation of older workers has anticipated their retirement as being no later than age 65. Governments have continually promised these workers that they would receive a government pension upon retirement. Indeed, as workers approach 65 in the United States—actually it's age 66 now for full Social Security eligibility—they receive a statement each year from the Social Security Administration showing them how much their retirement benefits would be at the age of full eligibility as well as their reduced benefit should they choose early retirement at age 62. To these workers, it is a government guarantee that these funds will be available once they reach the promised age of retirement.

Regardless, the retirement culture will have to change, and soon. What can be done to accelerate this change? Perhaps the best place to start is to look at how cultures have changed in the past. What brought about those changes and what lessons can we learn from them? Perhaps we can apply those lessons to bring about meaningful change in the current retirement environment.

## CULTURAL CHANGE AGENTS

The primary catalyst for rapid social change is war, which Americans have experienced firsthand in the past century with their involvement in World Wars I and II, Korea, Vietnam, and the wars in Iraq and Afghanistan. Each of these wars brought about a relatively sudden shift in the existing cultural paradigms of the time. Prior to the Vietnam War, for example, military service and U.S. military involvement in foreign wars was seen as a source of national pride. Vietnam brought about a rapid cultural change, but obviously resorting to war to change the retirement culture would be a little like clearing the clutter out of a room with a hand grenade. Another unthinkable possibility for rapid change would be a major pandemic, killing millions of the very old and very young. Certainly, these are events we would hope to avoid.

History shows us that major visionaries and political leaders (good and bad) have also dramatically transformed cultures. Martin Luther King Jr., Nelson Mandela, Mahatma Gandhi, Adolf Hitler, Karl Marx, Jesus, Mohammed—all of these individuals drastically changed their respective cultures, although in some cases, the respective movements would not reach full impact for many decades. Add to this the fact that cultural movements often lack a single leader, so it can be difficult to pinpoint one specific person as the catalyst for change. Examples include women's suffrage, minority voting rights, working women, and gay rights, as well as the rise of the current internet and computer cultures.

Looking elsewhere for clues to cultural change, economic hardship is a major change agent. The Great Depression of the 1930s played a key role in the passage of the Social Security Act and the continued economic malaise of Germany that eventually contributed to World War II. Historians might argue that without the economic collapse of Germany in the

1920s or the Great Depression of the 1930s, it would have been unlikely for Adolf Hitler to rise to power. Without that backdrop, no one would have cared about a young firebrand and his beer-hall polemic speeches.

In the past few years, we saw the role of economic hardship in the protests and unrest throughout Europe and the United States. As recently as 2010, as protests were breaking out in Greece and Spain over austerity measures designed to cut back out-of-control government spending, commentators scoffed at the possibility that similar protests could occur in the United States. Yet barely a year later in the summer of 2011, the Occupy Wall Street movement grew to become a worldwide movement in less than a year. In view of these examples, there is little doubt that economic hardship can be a powerful force in bringing about cultural change.

Money, of course, is also a major driving force for just about everything in modern society, so it stands to reason it would also be a factor in cultural change. Would modern retirement ever have become part of the American dream without Social Security pensions? In the 1960s, articles in popular magazines boasted that retirees could live in retirement communities for as little as $200 a month, less than their monthly Social Security benefit at the time.[3] Without old-age pension programs, the retirement culture would have probably developed along quite different lines than it did.

These early advertisements for living the good life in the retirement communities of half a century ago bring us to a very interesting observation. Savvy marketing teams helped create the current retirement structure. These same strategies could play a key role in reshaping the retirement culture in the future.

## CHANGING THE RETIREMENT CULTURE

Marketing in its broadest sense refers to selling not just products and services but also to selling ideas. Marketing has the power to change all types of cultural beliefs and has done so repeatedly throughout modern history.

The concept of retirement was created in the late 1800s by Otto von Bismarck, but it never really caught on as a major stage of life. The Industrial Revolution made it financially unattractive to keep older workers on the assembly line—where technology and money became cultural catalysts—and finally resulted in retirement being recognized as a stage of life, albeit a rather unpopular one. Old-age pensions and Medicare—examples of entitlement programs changing cultures—and the relative affluence of the working class after World War II made retirement possible for the masses, but even then it still wasn't widely accepted. Sure, there were some workers who looked forward to retirement, but the widespread cultural belief as late as 1960 was that retiring was being "put out to pasture." It wasn't until Del E. Webb began his massive marketing campaign for the Sun City retirement community in 1960 that it became a desirable stage of life and eventually a part of the American dream. Thus, marketing was the primary driving force in the creation of today's retirement culture, so it will probably play a key role in changing our perception of retirement in the future. If a major thought leader were to "market" a new perception of retirement (like Del E. Webb in 1960), that would be a tremendous asset, but the probability that one person will change the entire retirement culture is unrealistic.

Economic hardship is another way to market change. If the negative realities of the existing retirement mentality could be brought to the forefront of public consciousness, it could mobilize governments or institutions to change how they operate. The realities of economic hardship could be used as marketing tools to generate support for needed changes in Social Security laws.

Although there is bound to be an outcry when the retirement age is eventually changed, within a decade or so the new age will become accepted as the norm and the retirement culture will change accordingly. Unfortunately, this brings us to one of those catch-22s of life. Changing the age of eligibility for old-age pensions eventually changes the retirement culture, but it's very difficult to change the age of eligibility without first changing the retirement culture.

## LIFELONG LEARNING

As retirement changes, more seniors are going to be working than ever before. Physical limitations will not be the only obstacles to overcome. As technology increases at a breakneck pace, how will older workers stay up to date on the latest innovations? Many employers already believe seniors are technophobic, especially when compared to today's youth who grow up with computers and electronic gadgets. How will these seniors stay savvy enough to be productive in an increasingly electronic workplace?

There was a time, perhaps as recently as half a century ago, when people worked their entire lives in one career. For better or worse, those days are gone. Today, the average time of employment before switching to another job is about four years, according to the Bureau of Labor Statistics. It's not uncommon for people to have several careers throughout their working lives.

As they approach their retirement years, many workers become reluctant to try new things. That's unfortunate, because most older workers possess a wealth of experience, much of which they could carry over into a new profession. When children were at home, risk-taking was difficult, but once they leave the nest, starting a new career or even starting one's own business becomes more appealing. Without children, older workers can be more flexible when dealing with a fluctuating income, working hours, or the need to relocate for a new job opportunity.

In fact, the annals of history are filled with examples of senior entrepreneurs who became successful past the age when most people would be thinking of retirement. Harland Sanders ran a gas station with a restaurant until age 62, at which time he took the risk of franchising his idea to eventually become Kentucky Fried Chicken, opening more than 600 restaurants over the next ten years.[4] In 1863 at the age of 69—nearly 30 years older than the average life expectancy at the time—Cornelius Vanderbilt bought his first railroad.[5] Over the next two decades, he went on to own over a dozen railroads, founded Vanderbilt University, and became one of the wealthiest men in American history in GDP-adjusted dollars, second only to John D. Rockefeller.[6] Mary Kay Ash was a single mom at the age of 45 when she started her own business, which would eventually become Mary Kay Cosmetics. By 2008, Mary Kay Cosmetics boasted over 1.7 million consultants worldwide and revenue in excess of $2.2 billion.[7] In 1999, she was recognized as the Most Outstanding Woman in Business in the 20th Century by the Lifetime television network.[8]

The wisdom and experience that comes with advanced age can be used to great advantage by those willing to take on a challenge. John Astor was a German immigrant in the fledgling United States who became a successful realtor in the early 1800s. At the age of 71, he recognized that the next big growth opportunity would be the development of New York City. Even though he had no prior experience in real estate, he sold his fur trading business and other ventures in order to concentrate solely on buying land in Manhattan and the surrounding areas.[9] By the time of his death in 1848 at age 85, John Astor would be the richest man in the United States.[10]

With the advent of the Information Age, lifelong learning has become more important than ever. Consider the case of Red Jackson, a Michigan auto worker. Red worked on an assembly line of a major U.S. auto manufacturer for nearly three decades until he was laid off when his plant closed in 2012. Although Red worked with robotic devices for over a decade, his knowledge in the field was very narrow—he ran calibrations, monitored operational parameters, and so on. When Red was laid off, he had no useful skills outside of his previous job. He also had limited education to fall back on. For Red, this was a tragic personal situation, but on a larger scale, stories like this one are bad for the nation as a whole. Across the country, thousands of former workers like Red are not just unemployed; they place a growing burden on society as they draw unemployment benefits. Their reduced consumer spending further impacts a declining economy.

Computer skills would be a valuable asset to Red and other workers in his situation, but many older workers lack even the most basic computer knowledge. In spite of his admirable work ethic, Red would be unemployable in any information-oriented job because he lacks basic computer skills. Decades ago, an assembly line worker in Detroit could transfer his mechanical skills to other manufacturing jobs, but today most of those jobs have left the United States to regions with a more favorable business environment—lower taxes, cheaper labor, no union hassles, and a government infrastructure that provides tax benefits for relocating to their nation or region.

Even within the United States, many industries are relocating to more business-friendly states, leaving thousands of workers without jobs. Over the last decade, Nissan has relocated their national headquarters from California to Tennessee and established a new manufacturing facility for the Nissan Leaf in the town of Smyrna. Dell Computer built a new construction facility just east of Nashville. General Motors built a manufacturing facility in Spring Hill, Tennessee, for its now-discontinued Saturn line and plans to retool it for vehicle powertrains. This is good news for Tennessee and other Southern states gaining jobs, but it is bad news for Michigan and California, where unemployment is significantly higher because these laid-off workers lack the skills and education to pursue Information-Age careers. It's the same in other nations.

Take the case of Julian Stephanopoulos, a government employee in Greece who was laid off during the government's austerity movement. Like many Greeks, Julian prided himself on never learning English. This is typical of some Europeans who refuse to learn English out of a misguided sense of national pride. Since English is undoubtedly the international language of commerce, Julian's shortsightedness greatly limits his future employment opportunities.

What can be done to help these unemployed workers and other adults who need to keep up with a changing world? Part of the solution could be creating a structure that provides easier access to formal education programs designed to help workers of all ages develop the necessary skills to stay marketable in a rapidly changing job market.

## CREATING A LIFE-LONG LEARNING STRUCTURE

It would be relatively easy to establish an online structure to provide basic and even advanced educational classes for unemployed, underemployed, and aging workers. The curriculum could include ongoing career development in a wide range of fields at very little cost. Courses could be developed for business management, business finance, entrepreneurship, salesmanship, and of course, marketing. The technology to create this lifelong learning structure already exists, so there is no technological barrier to implementation. Funding would not be a significant problem. All that would be needed is a few dozen online interactive educational programs and an automated database to track eligibility, participation, and completion.

Traditionalists might argue that these courses would need teachers to respond to the tens of thousands or even millions of students who would have questions about some aspect of their studies, but an easy solution exists. These online universities could set up an online forum where students could post their questions and other students could answer them. Participants could then vote on the best answer. Similar forums, such as Yahoo! Answers, already exist. After a while, most of the common questions would have answers already posted. If there was a need for forum moderators, they could be solicited from previous qualified graduates, similar to the way graduate students are used in universities today. Serving as a forum moderator could then be a résumé enhancement for individuals seeking jobs. A library of short, focused videos could be developed by instructors to address more commonly asked questions and made available online through existing channels, such as YouTube.

An opportunity to generate revenue from these programs also exists. Companies could post internship opportunities online to be viewed by individuals signing up for certain electives. Job postings could be handled in a similar manner. Again this could be automated so that job postings would disappear in 30 days unless they were renewed by the prospective employer.

In short, there are many creative ways to make ongoing adult education affordable and practical. Using a completely automated format, the cost of providing this continuing education would be minimal.

Even the lack of a computer need not be a significant limiting factor. In 2011, an Indian company called Datawind announced that it would sell 100,000 tablet computers per month to their government to distribute to children in the classroom at a cost of $35 per tablet.[11] This tablet, the Aakash, is driven by Android 2.2 and has a color screen, two USB ports, and 256MB of RAM, and provides word processing, web browsing, and video conferencing. To put this in perspective, a similar limited-function tablet would cost less than one day's worth of unemployment compensation.

These continuing education programs would make laid-off workers and other lifetime learning adult students more employable and illustrate to potential employers their willingness to make themselves more marketable while looking for a new job. At the same time, if access to ongoing education programs reduced the amount of time that individuals were on unemployment, the savings could be significant. In 2012, over 5 million people were receiving unemployment benefits in the United States.[12] If unemployed workers got off the rolls even a few weeks earlier, it could save governments billions of dollars.

## CAREER DEVELOPMENT

Once education courses are established, the next logical step would be to incentivize ongoing education for workers who were already employed. This could be structured many ways. Obviously, a myriad of details need to be worked out and potential obstacles overcome prior to implementation, but that is always the case with any new program.

The core curriculum could include three courses: English for non-native speakers, information technology, and personal finance—knowledge of which is woefully absent from large segments of the U.S. and European workforces. Frankly, the core programs could be the same in the United States and Great Britain if English were replaced by remedial English. Large segments of the U.S. and U.K. populations either cannot speak English or possess poor reading and writing skills. Once these core programs are completed, individuals would be eligible to sign up for specific courses in other areas of interest, whether it's in one's current career or in some new career field.

Of course, there will always be some people who hate the idea of more education, just like there will always be kids who hate the idea of eating their vegetables. Human nature being what it is, it's hard to get some people to do anything even when it is in their own best interest. Still, lifelong learning has such tremendous benefits for individuals and for nations as a whole that it will probably be adopted by many countries in the coming years.

Businesses are likely to be strong supporters of lifelong learning because it would create a better educated workforce. Any government spending on continuing education would provide a tremendous return on investment by increasing worker productivity and making unemployed workers more employable. Studies show that nations with a higher-educated workforce have a higher level of prosperity and GDP per capita. It's a win-win for everyone.

In fact, college-level courses are already being provided online for free. Beginning in 2012, five prestigious U.S. universities—Stanford, Princeton, the University of California at Berkeley, the University of Pennsylvania, and the University of Michigan—collaborated to offer more than three dozen free online courses to students worldwide using videos from courses by professors at these prestigious universities. This exciting new educational platform is called Coursera (www.consera.org). In less than six months, the number of participating universities had grown to 16. By the end of 2012, participating universities had grown to 33, including such respected universities as Vanderbilt, Duke, Emory, Brown, Rice, and the University of London.[13] Coursera even embeds assignments and exams into video lectures and answers questions from students on online forums.[14] In its first year of operation, online attendance topped one million students.[15]

## SENIOR EDUCATION

It's not uncommon to see articles about people in their 90s who have gone back to college and gotten their degrees, but these stories are newsworthy precisely because they are the exception to the rule. In the future, it could be commonplace for seniors to go back to school to get advanced degrees to pursue a new vocation that is both personally challenging and fulfilling. It's true that work can be dreary, but it can also be tremendously fulfilling once the individual finds the job that suits his or her unique personality.

Take Josh, for example. Josh retired with his wife to New Smyrna Beach, Florida, after working on an assembly line in Michigan. Over the years, he grew to hate his job, but it paid well and gave him the opportunity to retire to a warmer climate. However, Josh quickly became bored. He always longed to be around boats and the water, so he volunteered to work at a local marina. After gaining some skills through on-the-job training, he was offered a paid part-time position. Josh now has the best of both worlds. On his working day, he gets to do a job that he loves and he meets interesting people, many of whom are very wealthy and own very expensive boats. On his days off, he often gets to go out on these yachts as a guest of his new friends or as a crew member. Josh's experience is not so different from that of millions of other retirees who leave the workforce only to discover that, after a while, full-time retirement can get very boring.

Josh had the right attitude to help him stay enthusiastic and involved with life, but millions of workers in the same situation lack the creativity or initiative to do this on their own. A lifelong learning program utilizing online education courses could give workers the training they need to move to new professions or to stay gainfully employed in their existing professions. This is, after all, the Information Age. It's time that information be widely and freely available or at least available at a nominal fee to the adult workforce. Studies show a strong correlation between national education level and economic prosperity. It's time to move education into the twenty-first century.

# ELEVEN
# Nature versus Nurture

*Reversing Psychological Aging*

At some point in life—sooner for some, later for others—the idea that we are getting older begins to creep into our thoughts, our attitudes, and our responses to casual remarks—it's not that we are just getting older, but that we are getting *old*. When approached with the opportunity to participate in an exciting new adventure, one senior responds with a negative affirmation we've all heard at one time or another, "I'm too old for that," while another exclaims with barely controlled excitement, "Let's go!" It's a different psychological perspective, to be sure, but what causes this difference?

Perhaps it is nature—hardwired into our brains for eons and passed on in our genes. That would seem likely, but if so, how do we explain the different responses just described? One of my business associates spends each winter traveling around the United States in a recreational vehicle. "The seniors are so different in RV parks," he explains. "They're active, involved, and always ready for a new adventure." Why are they so active when their peers are slowing down and withdrawing from the world? Obviously, there may be a certain amount of self-selection involved—active and outgoing people gravitate to an active and outgoing lifestyle. However, nurture cannot be ignored either. The cultural environment provided by fellow campers is bound to play a significant role. The lifestyle is nomadic and stimulating, which would seemingly lead to more adventurous behavior. Granted, touring the country in a $200,000 motor coach is a far cry from the risks faced by wagon trains in the Old West, but it is nonetheless more exciting than living alone.

On the other hand, these roving bands of gray-haired explorers could be genetic throwbacks—modern-day Ponce de Leons whom nature gifted with a genetic adventure switch permanently planted in the "on" position. When we look at these perpetually youthful road warriors, it begs the question, what causes our *attitudes* to age? The answer lies hidden within the gray matter of our brains in some area of human consciousness that science has yet to fully understand. Physical, biologic aging cannot fully account for all the marked differences that occur with age in the human species. To see the complete picture, we must also evaluate the unseen and the intangible—human psychology.

## PSYCHOLOGICAL AGING

Psychology plays a major role in aging. As we move through each stage of life, how we interact with others in the world around us is not just determined by our physical and mental capacity, but also by our psychological outlook on life, which progresses through stages as we move from childhood to maturity to old age. These stages have both nature and nurture components. Evolution teaches us that species change over time to adapt to the environment through a process of natural selection. As humans age, natural evolution favors ambition, risk-taking, and a forward-looking focus in early life, conservatism in midlife, and relative contentment in old age.

Risk-taking and ambition were positive evolutionary traits for early man. Paleolithic hunter-gatherers needed these characteristics just to survive when traversing dangerous terrain and braving the harsh elements. However, risk-taking was not just valuable in hunting prey but also in impressing potential mates. It allowed prehistoric man to survive to reach reproductive age, then venture out in search of a new partner (preferably one outside of his immediate family), then reproduce and care for their young. We can see similar behavior today—young men riding sport bikes at dangerously high speeds or indulging in other reckless endeavors, moving to the tune of an evolutionary dance that has endured for eons and contributed to the rise of modern civilization.

Some scientists believe that the relative contentment of old age and the reduction of risk-taking with age are products of our evolution. Others believe it is simply a byproduct of reduced testosterone with age. By the time a man reaches age 70, testosterone levels have dropped by 50 percent from the late teens. Either way, nature—whether in the form of evolution or biology—plays a significant role in psychological aging.

Nurture, on the other hand, also plays an important role in how we perceive ourselves with increasing age. In some cultures—notably Japanese, Chinese, Native American, and aboriginal Australian—the elderly are held in high esteem, but in most Western cultures, seniors are generally viewed negatively. This cultural bias against the old exerts a powerful subliminal message on seniors: "If everyone thinks I'm frail and over the hill," a senior might subconsciously rationalize, "perhaps that's how I should act."

Physical aging occurs so slowly in humans that once we reach adulthood we seldom notice the physical changes on a day-to-day or even year-to-year basis. But at some point, usually around middle age, our attitude about aging begins to change. Often, it's noticeable in some older adults who start to refer to their age when discussing almost any topic. "Let's go for a morning walk," someone says. "At my age, I feel lucky just to get up every morning." It's meant as a joke, but it also provides a window to the soul. Those individuals have begun to think of themselves as *old*. For many seniors, their attitude toward aging represents a psychological barrier that's just as real as any physical limitation.

## SOCIOEMOTIONAL SELECTIVITY THEORY

With advancing age, seniors tend to select social environments that provide positive emotional stimulus and avoid negative or neutral social interactions. This necessarily results in a narrowing of the individual's social circle. This phenomenon is called "Socioemotional Selectivity Theory." In extreme cases, seniors reaching advanced age seem to withdraw completely from society. It's as if they receive insufficient positive emotional stimulus from their contacts, so it's no longer worth the effort to stay involved. Although there are exceptions, these seniors tend to be less satisfied and content with life than their more socially involved peers. Meanwhile, psychologically healthier seniors retain a circle of friends and relatives who provide positive emotional feedback. To see examples, one needs look no further than grandparents interacting with grandchildren, especially very young grandchildren.

## FACTORS AFFECTING PSYCHOLOGICAL AGING

Psychological aging is a complex process that cannot be attributed to a single cause. Not only are there multiple causative factors, but the impact of these factors can vary greatly from one individual to another. Stress, depression, cultural expectations, and attitude can all play important roles. Of course, physical health plays a major role in psychological aging. Hormonal expression changes dramatically with age, rising fastest in teenage girls and then declining with age—abruptly for women in menopause and more gradually but no less intensely in men. Diminished mental and physical capacity as well as chronic exposure to pain from debilitating conditions can also have a negative impact on psychological health and well-being.

To some degree, physical aging and psychological aging go hand in hand, but there are many notable exceptions. I personally know many scientists who have begun to experience some of the debilitating effects of aging, yet they maintain a remarkably positive and youthful outlook on life. Why is it that some people seem to thrive on lifelong learning—college professors, for example—while others form relatively fixed views of the world? Why do some seem to dwell on the positive aspects of aging and others seem to be equally lost in the negative?

Rita is a case in point—one of those people who has seemingly given up on life at a relatively early age. She lives in a modest house trailer in an RV park in Lake Wales, Florida. Her husband died a few years ago, and the burden of maintaining two homes and driving back and forth between Florida and Michigan each year has begun to take its toll. In her conversations with visitors, Rita constantly refers to her inability to get around as well as she used to and the fact that she is "getting on in years." Rita made her home reasonably presentable and put it up for sale, but she laments that it is not nearly as attractive and well-kept as it was a few years ago.

From her physical appearance and especially her attitude, observers might reasonably guess Rita to be in her early 80s or at least a well-traveled late 70s, yet as her callers are saying goodbye, Rita reflects again on her advanced age, confessing that she will be 69 this spring. To Rita, that was very old, but she is actually one of the younger retirees in her park, and most have a far brighter outlook on life. Nature and nurture have taken a toll on Rita's psychological age.

Stress, such as that experienced by Rita starting with the loss of her husband, plays an important role in her rapid psychological aging. Individuals with poor stress-coping mechanisms can age faster and experience more health problems than those with relatively low stress and above-average coping skills. Culture also plays an important role. In many Western societies, the old are considered has-beens, useless and generally looked down upon by the rest of society. Derogatory terms like "geezer," "old fogey," "crone," "dinosaur," "curmudgeon," and "over the hill" imply that the old are no longer meaningful contributors to society.

Generally speaking, the more socially diverse and active seniors are, the younger they are psychologically. When our research team discussed Rita's situation with others in her community, we were not surprised to discover that she had an extremely limited circle of friends. Her next-door neighbor had never been inside her home, although they had been neighbors for over ten years and their homes sat less than 30 feet apart. In some ways, Rita is the poster child for negative aging. In would be very interesting to meet her circle of friends to see if their outlook on aging is similarly fatalistic. Meanwhile, Gladys lives in another retirement community less than ten miles away. Gladys also lost her husband years ago and suffers more significant physical disabilities than Rita, but her outlook on life remains unflappably positive. Not surprisingly, her social circle is wide and diverse—cards, bingo, potluck dinners, theater outings, day tours, and various social clubs keep Gladys so busy that relatives must work in their visits around her social schedule. If you spent any time talking with these two women, you'd think they were about the same age. In reality, Gladys is 91, more than 23 years older than Rita.

## REVERSING PSYCHOLOGICAL AGING

Many studies have been done on psychological aging, but relatively little research exists on how to reverse it. However, one approach that shows promise is the Selection, Optimization, and Compensation (SOC) theory, which suggests that humans compensate for declining mental and physical capacities by choosing their environments to get the most out of their diminished abilities.[1]

Sam is a retired corporate attorney from upstate New York who relocated to Florida a decade ago. While Sam has likely never heard of SOC theory, at age 75 he maintains an active social life that closely conforms to it. During his first years in Florida, Sam chose a series of social outlets that provided a positive atmosphere and emotional support. As a result, he continued to associate with like-minded individuals (including some fellow New Englanders) who had a similarly positive outlook on life and were involved in fulfilling activities.

To compensate for his reduced social circle in retirement, Sam optimized his involvement with these groups by developing strong friendships among a wide range of people. His new environment—a Florida retirement community where the pace of living was considerably slower than that of a big city attorney—struck the right balance between his need for socialization and his slightly declining energy and mental speed. As a result, Sam was able to interact easily with his peers and still seemed as sharp as ever to those who had known him for years in Florida (although those who knew him decades earlier as a practicing attorney might notice that his quick wit had slowed a step).

Sam subconsciously selected, optimized, and compensated for his declining physical abilities and reduced social circle in retirement. As a result, he leads an active and youthful retirement. Millions of retirees subconsciously follow Sam's approach when they retire. They select an environment that takes advantage of their strengths and minimizes their weaknesses. They optimize their social support structure by building stronger relationships than they had in the past.

Finally, successful retirees compensate for their declining abilities by creatively working around some of their limitations. Gladys, the 91-year-old widow mentioned earlier, has arranged for the weekly card games to be held at her home so she doesn't have to drive. She rides with neighbors to the bingo hall so she doesn't have to drive at night. She uses an electronic scooter to move around her retirement community, visiting friends and participating in its many neighborhood activities. These compensation skills allow her to maintain a very dynamic social life even though she has difficulty walking more than 100 feet without resting.

As a result, these two seniors embody SOC theory—they maintain a positive social environment and stay emotionally healthier than their peers who have gradually withdrawn from society. As working populations age, governments and businesses should consider utilizing SOC theory to optimize their aging workforce. An older worker might select a position that minimizes any loss of lost mental quickness by anticipating questions in department meetings and preparing responses to those questions in advance. Employers could provide useful agendas to assist in this process. Older workers might compensate for loss of energy by working flexible hours or fewer days per week.

So far, governments have shown little interest in reversing psychological aging, but as their populations continue to age, that will change. Governments will eventually realize that the financial benefits of reversing functional age cannot be fully realized unless psychological age is also pushed back. One big step governments could take to slow psychological aging would be to push the official retirement age much further into the future. Right now, retirement age is 60 in France, 65 in Germany and England, and 66 in the United States. These retirement ages send a strong message to workers that they are "expected" to retire at those ages. Not only is it expected, but it is their right. That message is further reinforced in media advertising using that age to market retirement services.

Although extending the retirement age to 70 might not have an immediate impact on psychological aging, within a few years (certainly within a decade) the new age would become the norm. Being "old" would not begin until age 70. Fortunately for individuals, the biggest single factor in psychological aging is one that we have not yet discussed—attitude. Studies consistently show that seniors with positive attitudes respond better to negative life events and are psychologically younger than their peers. Individuals who are prepared for possible negative life events are typically in a better position to weather them. As regenerative breakthroughs occur, it will take years or even decades for these interventions to reach widespread availability. Those in poor physical health may be too far gone to benefit from these eventual breakthroughs, so the longer we can stay healthy, the more likely we will be to benefit from major regenerative medicine breakthroughs as they become available. Regardless, good health and a positive attitude can slow both physical and psychological aging.

# TWELVE

# Preventive and Regenerative Medicine

Each year, billions of dollars are lost due to preventable health problems, not just in medical costs, but also in lost productivity. Regenerative medicine will usher in a new era of extreme longevity, but with that gift will come a huge personal responsibility to avoid costly medical conditions due to lifestyle choices and failure to follow preventive health guidelines. If you're overweight and can't find the time to exercise, if you start your day with two doughnuts and four cups of coffee, or if you're a busy mom who can't find time to schedule a routine breast cancer screening, you're one of millions of people who don't follow the recommended preventive steps that could provide early diagnosis for diseases or prevent them entirely.

The potential for financial savings from preventive medicine is significant. The Milken Institute, an independent economic think tank, looked at seven common chronic health problems (cancer, heart disease, diabetes, pulmonary disorders, mental disorders, hypertension, and stroke) to determine the financial drain they place on society. According to their study, businesses lose $1 trillion in productivity each year due to these conditions alone.[1] The study focused on reduced effectiveness on the job (presenteeism) and lost work days due to illness. Individual presenteeism alone cost employers $828 billion; lost workdays another $127 billion. In addition, the caregivers of ill family members cost employers another $80 billion in presenteeism and $11 billion due to lost work days. Imagine how much could be saved if simple preventive health measures would be taken by the average person.

More than half of all Americans do not receive many important prevention services, such as immunizations, screening tests for early detection of disease, and education about healthy habits and injury prevention in spite of the obvious personal benefits.[2] For example, half of U.S. women 40 and older do not get their annual mammograms even though they have insurance.[3] About one-third of adults aged 50 to 75 (22 million) are not up to date on their recommended colorectal screening.[4] Twenty-five million Americans have diabetes, and one-quarter of them are unaware of it, even though simple tests exist to identify this disease.[5]

It's not as if information doesn't already exist regarding the importance of recommended screenings. The United States Preventive Services Task Force has established comprehensive guidelines for a broad range of clinical preventive health-care services, such as screening, counseling, and preventive medications. These recommendations encompass screening services and treatments for the major health challenges at every stage of life, and include specific details for a number of diseases typically seen after the age of 50.

Yet in spite of the known benefit of early intervention, millions of people fail to comply with these suggestions. They're too busy, they naïvely believe a catastrophic illness can't happen to them, or they are afraid of what they might find out. But the impending financial crisis of Medicare will likely spur governments to make their preventive screening standards much stricter. A public service campaign might emphasize the value and importance of early screenings, for example, while offering incentives for prompt action, such as limited-time coverage of testing not typically covered under Medicare. Still, people are put off by the inconvenience of testing or they fear the discovery that they have a disease. Decades ago this fear would have been more understandable since a diagnosis of cancer, for example, was tantamount to a death sentence. Today, however, many forms of cancer are treatable if they can be diagnosed in the early stages.

Since preventive guidelines have already been established for many major illnesses, one way to increase compliance would be to attach a financial incentive to preventive screening. Much can be learned from the wellness industry that, in recent years, has studied what causes people to want to be more actively involved in their healthy choices. Seniors who complete required screenings, for example, could submit a form at the end of the year for a $50 rebate on their Medicare premiums.

Other strategies to increase compliance with preventive medicine guidelines can be found in the private sector. Many corporations encourage their employees to complete a Health Risk Assessment (HRA) annually, a questionnaire that collects data on health history, family history, and lifestyle. Then the resulting report provides an evaluation of the individual's current health status and future health risks along with recommendations to reduce these risks. Typically, the HRA is part of a more comprehensive prevention initiative that includes some combination of blood pressure screening, blood tests, urinalysis, fitness assessment, physician checkup, hearing and vision tests, and a consultation with a health promotion professional to discuss identified risk factors and establish a plan to address them.

It's not unreasonable to predict that job applicants would face the following scenario in the not-too-distant future where the value of preventive medicine services has finally been recognized. First, the applicant would be required to complete an HRA to qualify for the company's health insurance plan as part of the initial paperwork. Later the applicant is asked to provide samples for blood work, genetic testing, and urinalysis. When the preliminary work-up is accomplished, an appointment with the company's health promotion counselor is scheduled. Together, they review family history, genetic markers, previous health history, and current lifestyle to determine what proactive steps the applicant should take to stay as healthy as possible, given risk factors and test results.

In this hypothetical case, the applicant discovers that, because of a family history of Alzheimer's and the presence of the APOE4 gene in her genetic test, there exists a four-times-greater predisposition of eventually getting the disease than the general population. And although tests also come up negative for diabetes, they discover a prediabetic condition. Since 11 percent of prediabetics become diabetics each year, there is a very strong likelihood that without preventive measures, diabetes will manifest itself within the next ten years.[6] Given that diabetes also increases the risk of Alzheimer's disease—not to mention doubling the risk of breast cancer—the applicant and her counselor agree that she needs to start preventive steps now to avoid these three diseases. The applicant is also about 60 pounds overweight, so two obvious steps are diet and exercise.

After the applicant is hired, her health promotion counselor recommends a nutritionist whose services are paid for by the employer and the insurance company. The counselor also provides a reference to a personal trainer, who assists in outlining an exercise program appropriate for age, gender, and specific needs. The trainer puts her in contact with a support group of people training toward similar health goals. Enrolling in a local fitness center and deducting enrollment fees from her income taxes is also an option.

Since the employer also picks up part of her fitness center dues, she decides this is the best choice for her. She quickly finds a new group of like-minded friends who are also determined to make their personal health a top priority. She is soon swapping healthy recipes and training with a group of women to complete a marathon. A year later, the employee has lost 60 pounds, and has successfully completed the Chicago Marathon. One simple questionnaire changed her life. In the end, of course, it was dedication and motivation that made it happen, but the proactive cultural and societal support the employee received along the way was the catalyst that got her off the sofa, into the fitness center, and eventually into her first marathon. But it all started with an HRA, health screening, and availability of supportive lifestyle services.

In the real world, HRAs have been shown to provide actual reduction in health-care expenses even when there is no follow-up counseling. Apparently, just raising awareness of one's health risk is enough to encourage behavioral change in some employees. Under current U.S. law, employers cannot mandate Health Risk Appraisals, but they can provide incentives to complete the questionnaire, such as a reduction in health insurance premiums or a cash incentive, typically between $25 and $100.

Each year, thousands of people decide they've had enough of seeing their health move in a downhill spiral toward illness and poor quality of life. They join a fitness center; they join a walking club; they start taking charge of their own health. Sadly, a far greater number of people think about making similar changes, but never do so because they lack the catalyst to get started or the incentive to stay with their wellness program. A coordinated national prevention program backed up by the will of motivated professionals could turn around this neglect of personal responsibility for one's own health.

## REGENERATIVE MEDICINE SCREENING

In the near future, it's likely that regenerative medicine will result in a new wave of preventive screenings. In some cases, it might be as simple as drawing an extra vial of blood to test for proteins associated with specific types of cancer. In other cases, genetic screenings might identify one's predisposition to specific types of cancer.

Championing this movement will be a new generation of philanthropists, led by entrepreneurs like Sergey Brin, cofounder of Google. In chapter one, we learned how Brin's wife, Anne Wojcicki, launched a company called 23andMe. The name refers to the 23 chromosome pairs found in all humans. The company provides partial DNA sequencing for just $99. Brin served as an alpha tester on the project. Although his initial screening showed nothing unusual, Brin's mother and an aunt had been diagnosed with Parkinson's disease at about the same age. His mother, who had also been tested by 23andMe, had a mutation to the LRRK2 gene, so Wojcicki suggested testing her husband for it as well. It turns out that he did share this rare mutation, which significantly increases the risk of Parkinson's disease.[7] The mutation does not mean that the carrier will develop Parkinson's, but it increases the probability to over 30 percent. Rather than take this prognosis passively, Brin began his own extensive research into how to prevent Parkinson's disease. He converted to a strict diet and exercise regimen and subsequently donated $50 million to Parkinson's research. In the past, many wealthy benefactors have donated millions to research diseases that had already afflicted either themselves or a family member, but Brin is believed to be the first person to fund scientific research in the hopes of preventing a disease for which the individual held a genetic predisposition.

Brin is also using the considerable resources available to him through the Google organization to accelerate the pace of medical research in Parkinson's using a process called data mining. Rather than start with a specific hypothesis and a very narrow research objective, data mining involves starting with a massive database of seemingly unrelated data and searching for possible correlations. The initial results are so promising that the practice could literally change the future of medical research.

Recently, the *New England Journal of Medicine* published the results of a massive worldwide study that identified a possible connection between Gaucher's disease—an excessive fatty buildup in internal organs—with a greatly increased risk of Alzheimer's disease. William Langston, director of the Parkinson's Institute, called 23andMe scientist Nicholas Eriksson and asked him to search for a similar correlation in 23andMe's customer database, which at the time already numbered more than 100,000. In about 20 minutes, Eriksson was able to calculate that Parkinson's sufferers were five times more likely to carry the Gaucher mutation than the general population. This has been confirmed in the *New England Journal of Medicine*'s own study as well.[8] Whereas traditional research had taken years and many millions of dollars to expose this link, data mining had found it within a matter of minutes. Its implications for future medical research are monumental. Once enough people have completed DNA sequencing and enough computing power is utilized, researchers will have the ability to search for previously unknown patterns between diseases that could accelerate the search for cures and preventive care.

Eventually, a simple blood test will be able to sequence the entire human genome, but the cost will most likely remain prohibitively high for routine testing in the foreseeable future. The first human genome was sequenced in 1998 at an estimated 13-year cumulative cost of about $3 billion.[9] As recently as 2007, the cost still stood at $10 million, but it has subsequently been falling precipitously, dropping below $10,000 by 2011.[10] In 2012, Life Technologies Inc. announced a new machine that could potentially sequence complete human genomes for about $1,000.[11]

However, it's not necessary to sequence the entire genome—99.9 percent of the genome is identical from one person to another—and this sequencing can be done at less cost. The sections of DNA that vary are called SNPs, or single-nucleotide polymorphisms. Companies, such as 23andMe, can already sequence hundreds of thousands of SNPs. Although this is only a small fraction of the 10 million SNPs in human DNA, many of these are "tag" SNPs, meaning they provide data on a group of related SNPs. This maximizes the information from each SNP analyzed, while keeping the cost low. It's possible that partial DNA testing could be conducted as part of a routine blood panel in the near future using techniques similar to those of 23andMe.

From a medical expense perspective, DNA sequencing promises answers to some of medicine's most perplexing questions. Why can some people smoke their entire lives and never get cancer? Why do a few lucky people live to advanced age with virtually no major health problems? The answer lies in their genes, hidden until now, but soon to be revealed. If we can discover what genes prevent lung cancer in smokers, perhaps we can discover a way to turn those genes off or on to achieve the same result in all smokers. If a gene, or group of genes, causes the food cravings that lead to obesity in many people, perhaps those genes can be controlled in the future as well. The fault is not in the stars, but in ourselves, as Cassius said to Brutus. Poetic indeed—but if *faults in our genes* predispose us to disease and death, that's even more rationale to increase regenerative medicine research.

## HEALTH-CARE REFORM

Even as remarkable anti-aging breakthroughs begin to appear in the near future, we must face the fact that such advances will take several years before becoming sufficiently widespread to have an impact on health-care expenses. This will be especially true for the expenses of millions of seniors who are already experiencing a multitude of age-related conditions. And although no one denies the need for broad health insurance reform in the United States, since the coming economic crisis will be driven by Medicare expenses, this chapter focuses solely on the inherent problems and challenges of the program as it exists today. The first and most obvious issue is the declining worker to senior ratio, which means less and less funds feeding in to the system as more and more people are drawing from it.

The problems of Medicare are complex. Besides the declining ratio of workers to seniors, the day-to-day operations of the program are expensive.

### ADMINISTRATION:

About six cents of every Medicare dollar goes directly toward administration of the Medicare program, but beyond that expenses include the administrative costs of the hospitals and doctors that provide the actual care. A more realistic estimate might be 25 percent that goes to administration in total.[12] One area that could potentially result in substantial savings is by automating patient records, which would not only reduce administrative expenses, but also prevent dangerous side effects from drug interactions and ensure timely emergent or urgent care regardless of where the senior might be when it is needed. While there is strong support for automation both in the private and public sector, the likelihood of a system-wide network is a number of years off.

### MALPRACTICE INSURANCE:

This type of insurance adds an estimated 1.5 percent to health-care costs, or about $30 billion a year of U.S. health care's $2 trillion cost. Although

1.5 percent may seem trivial, $30 billion per year is about equal to the entire NIH research funding budget and ten times more than the total NIH funding for aging research.[13] Granted, sometimes litigation is warranted, but meritless litigation takes a heavy toll on the morale of physicians who must deal with the emotional and reputational damage. As a result, some physicians are leaving medicine or moving to specialties with less expensive insurance. A study by Merritt Hawkins and the Physicians' Foundation found that one in ten physicians planned to leave health care in the next three years.[14] If this trend continues, by 2020 the health-care system will have a shortage of 40,000 primary care doctors.[15]

Fear of litigation also leads to defensive medicine: unnecessary and usually expensive tests and procedures in order to avoid lawsuits. A doctor might order three tests at once, for example, when a positive result from the first test might rule out the need for the expense of additional testing. A doctor might not wait to order tests when the most prudent approach would be to wait and see if the body healed itself naturally. The total cost of these and similar defensive medicine practices has been estimated to be as high as $210 billion each year.[16]

## PHARMACEUTICAL INDUSTRY REFORM:

See chapter 9 for a discussion on the problems of the pharmaceutical industry. In short, the pharmaceutical industry makes huge contributions to both political parties and is represented by an extremely effective lobby. Reforming the pharmaceutical industry would be a formidable challenge.

## CLINICAL TRIALS AND INFORMED CONSENT:

Clinical trials can often take many years to complete, during which time doctors cannot prescribe the experimental drugs to patients, even those who may be terminally ill. In such cases, the Hippocratic mandate of "do no harm" translates into a death sentence. Given a choice between waiting to die or trying an experimental drug before it completes the rigorous FDA approval process, many terminally ill patients would choose the latter. If patients and their immediate heirs were allowed to sign an "informed consent" form that releases doctors and the hospital from litigation should the patient die or suffer other ill effects from the drug, valuable trials could move forward at a much faster pace to the benefit of all mankind.

Informed consent could eventually expand beyond terminal patients to provide substantial savings to the Medicare and Medicaid systems. As a case in point, the prescription drug bexarotene has been approved for over ten years to treat a certain type of lymphoma, but recently researchers at the Case Western University Hospital found that it could dramatically and rapidly reverse Alzheimer's disease in mice.[17] The drug apparently works by breaking up amyloid plaques that build up outside brain cells and interfere with cognitive function. Researchers found that within 72 hours the amount of amyloid plaque decreased by 75 percent and mice returned to normal nest-building behaviors when pieces of tissue paper were introduced into their cage, indicating a return to normal brain functioning.[18]

As a result of these promising findings, bexarotene will enter clinical trials for Alzheimer's disease on the FDA "fast-track" program and might be approved for Alzheimer's patients in a couple of years. Meanwhile, 5.4 million people in the United States alone have Alzheimer's disease with combined health-care costs of over $200 billion annually.[19] Why should Alzheimer's patients be forced to wait a year or more to obtain this drug? It's been approved for cancer treatment for over ten years so its side effects are well-known—most symptoms are similar to those of the common flu. This is a good example of how well-meaning policies can become bureaucratic red tape. Patients should be allowed to weigh all available information and make informed decisions on the use of experimental drugs and medical devices.

### WELLNESS COMPLIANCE:

Currently individuals who abuse their health receive the same standard of care and the same Medicare coverage as those who religiously follow their doctors' recommendations for managing existing illnesses or staying as healthy as possible. For example, obesity increases Medicare expenditures by about two-thirds, yet the obese pay the same Medicare premiums and have access to the same level of medical care as those who have taken steps to maintain their weight at healthy levels.[20] Patients who refuse to comply with their doctors' instructions for rehabilitation often incur additional medical expenses under the current system, yet they face no financial consequences from their failure to comply with rehabilitation protocols. Their premiums stay the same, they incur no risk of losing Medicare coverage, and they continue to receive the same standard of care for related future complications as those patients who responsibly comply with the rehabilitation protocols *even when their lack of compliance was the cause of the complication.*

Imagine the chaos that would ensue if businesses followed the same policy. If an employee with a mission-critical task forgets to perform it every day, they may be counseled or reprimanded the first and maybe even the second time it happens, but if the conduct continued, the employee would be fired. *In our current health-care system, there are no financial consequences for bad behavior.*

There is, actually, an emerging trend to reward preventive steps taken by employees. In some states, businesses require obese employees to either show weight loss or participate in weight-management programs to qualify for discounts on their health insurance premiums. If they fail to comply, they must pay premiums at the undiscounted rate, which is tantamount to an obesity surcharge. From the business perspective, this is a good financial move because it reduces productivity loss and the cost of future health insurance premiums paid by the employer. Frankly, it is also in the best interests of the employee as well.

Japan is doing something similar, requiring that all employees meet certain waist size standards. Those employees who do not meet the standard and don't have a medical excuse are given guidance on how to reduce weight. Companies that fail to reach government-established targets face stiff penalties. This law, called the "metabo law" in Japan, went into force in 2008. So far, it has resulted in significant behavioral changes in the targeted population. Although it is definitely controversial, it's not too far-fetched to imagine other nations following Japan's example as they attempt to rein in the rising costs of health care.

## LAST MILE ALTERNATIVES:

Approximately 28 percent of Medicare expenses occur over the last year of life, or about $150 billion each year.[21] That's larger than the budget for the entire Department of Education or the Department of Homeland Security. About 12 percent occurs over the last two months of life.[22] Of course, one would logically expect expenses to be greater during this period, but experts believe that 20 to 30 percent of these expenses have very little impact on the patient's life expectancy or quality-of-life over their remaining days.[23]

In an article titled "How Doctors Die," Ken Murray reveals that doctors don't die like the rest of us.[24] They have seen the pain inflicted on dying patients by well-meaning but ignorant family members who don't understand the limits of modern medicine, and they don't want to go through that themselves. And they know enough about death to know what all people fear most: dying in pain, and dying alone. Consequently, when faced with heroic measures to maintain life, doctors overwhelmingly opt for a more conservative approach, choosing to die without heroic measures to maintain life.

The current system encourages overtreatment in the end stage of life. Often, overwhelmed family members request that doctors do everything to save a loved one, but all these patients gain in their final moments is a flurry of poking, moving, and frightening visions of medical personnel scurrying around. Instead, these dying patients could be quietly and serenely visiting with loving family awaiting the end. It's not pleasant to think about—death never is—but when observing how most people die, alone and connected to machines and tubes, most doctors opt for a more humane approach when facing their final hours. More of us should follow that example.

An associate of mine relates this story. His elderly mother was dying from numerous conditions—so many that it was mere guesswork to determine which one would ultimately be the cause of death. When doctors recommended a battery of tests—requiring that she leave the relative comfort of her bed for an ambulance ride to the hospital and a day of poking and probing by strangers, he asked what benefits the tests would be for his mother. When they admitted that her care would continue as before regardless of the outcome of the test, he flatly refused to allow it, saying, "I wouldn't treat a dying pet that inhumanely, and I certainly won't do it with my mother."[25] Instead, the adult children sat with the parent until her last breath.

Of course, people have strong opinions on this issue. They'll want extreme measures and they are entitled to them. What's really needed is a civil discussion to raise awareness among the dying and the survivors of the best options in these last days of life. It's not just a matter of dollars; it's a matter of compassion and allowing those who have chosen a more natural ending to die in dignity.

## MEDICARE TRIAGE

On the battlefield, military physicians face horrific life-and-death decisions. With limited resources and a small number of surgeons, doctors must make difficult choices on how to achieve the maximum benefit. Triage is a process whereby the level of intervention is determined by the limitations of available resources and the severity of the condition. It was first utilized by French doctors during World War I to handle battlefield injuries when the number of patients exceeded the limits of available care. The basic triage concept separates patients into three categories: 1) those who are likely to live regardless of the care received; 2) those who are likely to die regardless of the care received; and 3) those for whom immediate care might make the difference between life and death. Over time, the triage process has become more sophisticated and is now widely used in emergency rooms and in disaster situations.

Under the current Medicare system, however, no attempts of intervention are spared if the patient desires them, regardless of the futility of the eventual outcome. A *New York Times* article outlined the final days of Terence Foley, as chronicled by his wife, Amanda Bennett.[26] During this time, Medicare expenditures per day were higher than the typical senior's Medicare expenses for an entire year. On the one hand, these heroic attempts to preserve life without regard to cost represent the highest levels of human compassion. On the other, they illustrate humankind's inability to accept death.

As a result of these heroic efforts to preserve life, Medicare expenses over the last year of life average 28 percent of lifetime Medicare expenses.[27] The current approach to treating the dying is neither compassionate nor financially viable in today's economic reality. About one in five seniors spend their last few days in the ICU (intensive care unit) where the costs typically average $2000–$3000 per day, and upward of $10,000 in some hospitals. These patients—connected to tubes and often drugged out of their minds—die in relative isolation, separated from loved ones due to the restrictions of intensive care. Given the opportunity to look back on those final days, few dying seniors would choose such a cold, sterile environment over the warmth and compassion of a hospice or even their own home. There, they could receive palliative care as needed and be among their beloved friends, relatives, and cherished pets as the inevitable end approaches. Few attempts have been made to meaningfully address end-of-life expenses, but that could be changed with a campaign to make the public more aware of compassionate alternatives and the consequences of their end-of-life choices.

## PUBLIC AWARENESS CAMPAIGN:

Most people have no idea how much end-of-life expenses cost. Frankly, they have very little idea on how much any medical treatments cost unless the expense is disallowed for some procedural reason and they get an enormous bill. A tastefully presented series of public service advertisements could raise awareness about these unnecessary expenditures and how they are increasing the cost of Medicare for everyone. These ads could also point out that invasive and painful interventions might not be what their loved ones want or how they would choose to spend their remaining days. Most are not aware that many dying patients in ICU must be heavily sedated to prevent them from fighting to remove breathing tubes or IVs. To put it bluntly, it's not a very humane way to die, yet this is how modern medicine approaches death. In reality, there are many other options that allow for a more compassionate ending. Patients can be transferred to the more home-like surroundings of a hospice where they receive constant attention, painkillers as needed, and emotional support for both the dying and the bereaved survivors.

A friend of mine tells this story. "My mother was 91," he recalled. "She had been in declining health for some time and was living in a nursing home when she had a stroke. The relatives had been called in from around the country and were now waiting for the inevitable. On one of my visits to the ICU, I asked my mother how she felt. In a weak voice she replied, 'I hurt.' 'Where does it hurt?' I asked reflexively, realizing there was little I could do. 'My lips,' she replied. 'Your lips?' She nodded weakly. 'Are your lips chapped?' I asked."

Eyes closed, his mother nodded. He reached into his pocket and pulled out a tube of Chapstick and put some on his mother's lips. "She smiled at me," he recalled.

Such is the state of modern end-of-life care. Hospitals and doctors excel in providing the latest state-of-the-art in high-tech medicine, but very little human touch. In end-of-life medicine, it's really about compassion and reaching out. Sometimes, it's just a matter of being there to adjust a pillow or apply Chapstick. Compassion is what patients and their loved ones need at the end of life.

Public service advertisements on television and radio could urge couples to have end-of-life discussions with family, ensuring that everyone understands what level of care is desired and that interventions do not go beyond that. Standard living wills and advanced medical directives should be promoted in these public service announcements.

## EDUCATE THE PUBLIC ON MEDICAL TERMINOLOGY:

Many people simply don't understand medical terminology. In the future, medical professionals will most likely be trained to better communicate with patients and families in terms they can understand. For example, many people don't understand what a do-not-resuscitate (DNR) directive involves and, as a result, fail to agree to it. A DNR is a type of advanced *directive*, agreed to by the patient and signed by a physician which defines what should (or shouldn't) be done to a patient in the event of an emergency. Typically this means that if patients stop breathing or the heart stops beating, the medical team does not try to aggressively revive them. It's a humane measure that usually applies to elderly or terminal patients.

## ADVANCED DIRECTIVES:

Currently, living wills and advanced medical directives are discussed upon hospital or nursing home admission, but these are highly emotional times for the patient and immediate family. In the future, these medical care documents might be treated more like a will, discussing them as needed with immediate family and keeping them up to date as the circumstances change. They should also be brought to the family's attention when certain milestones—such as a specified decline in health—are reached.

## WIDER USE OF HOSPICE:

Hospices are underutilized. Often, terminal patients spend less than a week in hospice when their fate has been known several weeks or months in advance. Under the current system, it's difficult to get into hospice because the patient must first be declared as terminal, and as we have discussed, doctors are reluctant to take this step.

We're already seeing an increase in the use of hospices, but I have spoken to families who wanted to transfer a loved one there, but were unable to do so because they were unable to obtain a definitive terminal diagnosis. In the future, it's likely that hospice care will be made available earlier if the patient and immediate family sign a release requesting only palliative care and watchful waiting for the patient's remaining days.

## FORCED REDUCED STANDARD OF CARE:

Eventually, societies may take even more drastic measures to manage end-of-life expenses, such as changing the standard of care for terminal patients or patients with multiple illnesses deemed not able to benefit from extensive treatments. China would most likely be the country to do this first since its government permits unilateral changes in health care without regard to its approval by citizens.

One example of this situation would be when an elderly patient has congestive heart failure. Although this disease can be treated in younger patients, the surgery itself is currently very invasive and could potentially be fatal. Once seniors reach advanced age with congestive heart failure, surgery is no longer a viable option, so the disease is managed for the patient's remaining days. Eventually the patient either dies from heart failure or from the damage caused by reduction in blood flow to vital organs. For seniors who choose to quietly wait out their remaining days in a nursing home or hospice, this is not so bad from a financial perspective, but for patients who have multiple conditions that are being aggressively treated, the subsequent costs can be enormous. An extended stay in ICU can cost over $2,000 per day.[28]

## AGING RESEARCH FUNDING

As you can see, the wasted health-care funding discussed here cumulatively costs hundreds of billions of dollars. The money wasted in just one of these areas is typically greater than the entire budget for aging research. It's essential that aging research be given a higher funding priority in the future. Our economic stability depends on it.

# THIRTEEN
# Preparing for the Future

**T**he world is facing two widely divergent future scenarios. On one side, medical advances could greatly postpone old age, leading to extreme longevity that could render the problems created by old-age welfare programs irrelevant. On the other, continued deficit spending on senior entitlement programs could lead to hyperinflation, depression, stock market crashes, severe economic hardship, civil unrest, or even the toppling of governments. Planning for two such diverse future scenarios

## Longer Lives: Source of Growth or Decline

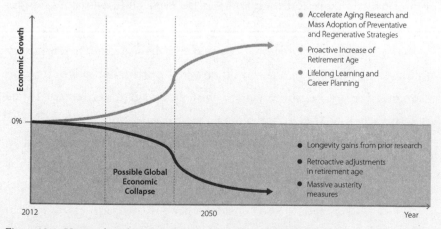

Figure 13.1. Unprecedented economic growth or lengthy depression.

might seem to be almost impossible, yet there are steps you can take to make a meaningful difference regardless of how the future unfolds.

To assist your planning, let's break down recommended actions into four categories: personal finance, health, life choices, and getting involved.

### PERSONAL FINANCE

The foundation of personal finance is risk management, so in order to weather the coming storm, you need to be prepared financially for the most likely outcomes. First, it's likely that you will have to pay substantially more out-of-pocket health-care expenses than seniors in the past. According to the Employee Benefits Research Institute (EBRI), a typical couple would need to save $158,000 by retirement just to have a 50/50 chance of funding their out-of-pocket health-care expenses over their remaining life expectancy.[1] To increase the odds to 90 percent, they would have to save $271,000.[2] However, that figure does not include the possibility of an extended stay in a nursing home, which could raise the number even higher.

Another possibility that could increase out-of-pocket health-care costs for seniors is a rise in Medicare premiums. One very likely possibility would be increase Medicare premiums for the wealthy in stages based on income. This is called means testing; it already exists for Medicare premiums, but it doesn't kick in until very high incomes. In the future, politicians will likely add more brackets so even middle-income retirees will pay higher Medicare premiums.

Since long-term care is only covered by Medicare under temporary circumstances, older workers should consider purchasing long-term-care insurance. Otherwise, those with seemingly adequate retirement funds could find their savings decimated by a few years in a nursing home. The downside is that long-term care insurance is expensive and complex, so it's best to work with a financial professional to determine what is best for your situation.

Older workers should also consider a patient advocacy service to help cut through the red tape when medical care becomes necessary. These telephone-based services can be obtained on short notice on an as-needed basis by paying an hourly rate. These advocates can also be retained by the adult children of aging seniors to help monitor the health-care of a parent in another state, potentially improving care and saving thousands of dollars in out-of-pocket medical expenses. For example, it's often just a matter of how a condition is coded in medical records that determines whether or not it's covered by Medicare. In some cases, a patient's response to a question might result in the procedure being disallowed for Medicare reimbursement whereas another equally truthful statement might result in the procedure being covered. It's impossible for the layperson to understand the nuances and intricacies of the health-care system, making the services of a patient advocate invaluable for achieving the best possible patient outcome.

Aging workers who lack sufficient funds to retire will have no choice but to continue working far past age 65. For all the reasons we've previously discussed, even those who think they have adequate savings will probably need substantially more funds than they've anticipated. Unless you're very wealthy, working past age 65 will be unavoidable, so you need to prepare accordingly. It's important to maintain good health, not only to maintain one's earnings ability after age 65, but also to remain employable. Workers who are laid off will have an extremely difficult time finding other employment if they have substantial health problems. Obesity alone would make it difficult for an aging worker to find a new job, regardless of the anti-obesity legislation that might be in place at the time. It's also important for workers to maintain employable skills, especially skills that would be in demand during a recession or economic slowdown.

It's also important to arrange personal finances to reduce any debt. Except for a few countries, notably Japan, which has a very high savings rate, most citizens of developed nations are overspending and undersaving. In the United States, for example, where most automobile purchases are financed, it's not uncommon for a couple to have combined car payments in excess of $10,000 a year. If that couple has earnings greater than $60,555 (28 percent tax bracket), they would need to earn over $15,000 to pay for those car payments.[3] It makes sense to reduce spending in order to live within your means and set aside sufficient savings for an emergency.

## HEALTH

Many books that provide investment advice to prepare for future uncertainties fail to mention your most important financial assets: your own health and your family's health. Maintaining good health reduces the risk of future medical expenses, but it also does something arguably even more important. It protects your human capital—the ability to earn income to offset stock market declines or other unforeseen financial pitfalls. Without good health, all other financial advantages become almost meaningless. Thomas Jefferson, a strong advocate of exercise and pursuit of good health, wrote, "Health is the first of all objects."[4]

At a recent international aging conference, a gerontologist gave this simple health advice: "Do what your mother told you." Typically, this translates into the following commands: eat your fruits and vegetables, don't overeat, don't smoke, and get enough sleep. Add to that simple list an admonition to exercise regularly, avoid stress, take a high-quality vitamin/mineral supplement, and avoid environmental pollution. Following these healthy lifestyle recommendations combined with an intellectually stimulating job will add several years to your statistical life expectancy, giving you many more years of healthy, productive life.

Good diet and exercise will not help you slow aging and extend healthy life significantly, but these simple and proven strategies may help you live long enough to take advantage of even more advanced regenerative strategies as they are discovered. In the past 50 years, the world has changed beyond recognition, and it is very likely that biomedical advances will progress from the lab to the clinic faster than in the past. Some of these methods may extend life expectancy by many years, buying time to develop even more life-extending procedures.

Many of my older friends are already taking proactive steps to maintain good health by taking drugs for their anti-aging benefits. These drugs include statins, beta blockers, ace inhibitors, anticlotting drugs, low-dose aspirin, multivitamins, and other drugs and supplements that are either prescribed by their physicians or available over-the-counter in their respective countries.

Statins, for example, lower the risk of cardiovascular disease, but they also appear to have other positive anti-aging benefits, such as reducing the risk of cancer and plaque buildup in arterial walls.[5] Using drugs for benefits outside their recommended protocol is called off-label use, but expanding the use of statins for these preventive benefits would provide substantial benefits to seniors.[6] It would also provide a massive financial benefit for pharmaceutical companies, as statins are already one of the most widely prescribed drugs with over $20 billion in annual sales.[7]

Since the profit potential is so great (and pharmaceutical companies have so much influence), it's very likely that these off-label uses for major prescription drugs will be researched and fast-tracked through the FDA approval process, but until then, it's not likely that you will hear about these anti-aging benefits from your primary care physician. Most doctors will not prescribe a drug unless you have a condition that is treatable by that drug. Doctors are bound by the Hippocratic Oath to "do no harm," and of course the possibility of litigation should any side effects occur. This doesn't mean you should rush to your doctor (or your corner drugstore if you live in a country where such drugs are readily available over the counter) to buy statins or other drugs for off-label use, but it doesn't hurt to make yourself aware of recent trends in the pharmaceutical world and form your own opinion.

Naturally, there is always a school of thought that says we should wait until a massive clinical study is performed on younger adults before taking any compounds designed to treat conditions that primarily occur in the older population, even though such studies might take decades to perform. Whether you choose the wait-and-see approach or a more aggressive stance is up to you. The primary reason it's being discussed here is to make you aware that this choice even exists. Just because your current doctor doesn't suggest it when you go in for a checkup does not mean that you couldn't benefit from a more proactive approach in communicating with your physician about your own personal and future health.

At a very minimum, it's critically important to maintain the highest level of health you can personally attain. For most people, this means losing weight, exercising more, eating right, and taking a far more proactive stance on preventive medicine and checkups. Further, be a role model and encourage your spouse (and children and parents) to take all these steps as well.

Becoming an informed consumer of health-care services is also a prudent step, especially if you have a condition that is likely to be chronic or worsen with age. Most of all, keep an open mind. Alternative medicines don't work for everyone, but a particular approach might work for someone of your particular genetic makeup. Don't summarily disregard all alternative medicines just because you tried something else a few years back and it didn't work.

## THE CASE FOR A PROACTIVE MINDSET

In order to benefit from coming anti-aging breakthroughs, it's absolutely essential to keep yourself in as good a state of health as possible so you will have the opportunity to benefit from those breakthroughs. It could literally mean the difference between life and death. Regenerative medicine breakthroughs are not likely to occur at the same pace in all areas. If a cure for cancer is found in the next five years, it won't help someone who is dying from chronic heart disease, so it's important to take all reasonable steps to optimize your health as much as possible. Thus, an open mind and a proactive mindset are absolutely essential to your future health. To illustrate this, consider antivirals and influenza.

The fight between humankind and viruses has been going on for millennia, and by looking at recent population growth statistics, we are winning. In 1918 (although the date varies by nation), the Great Flu Epidemic killed up to one-third of the population in some countries. Fortunately, most strains of influenza are less virulent. All of us have had the flu from time to time; many of us get it every winter. A decade ago, most doctors would only recommend hot tea, lots of fluid, and perhaps aspirin for the aches and pains. Even now, it's only after the virus progresses to the bacterial stage that doctors will consider prescribing antibiotics.

As a result, the typical sufferer approaches the flu with a level of passive resignation. Once the telltale signs start, most believe there is little that can be done. What they don't realize is that over time these frequent bouts with the flu can have a degenerative effect on the aging process and on the immune system in particular.

Before the swine flu (or H1N1) scare in 2009, I didn't know many potent ways to fight the influenza virus existed. As you might recall, during that outbreak, it was almost impossible to obtain Tamiflu (oseltamivir) in the United States because panicked consumers had demanded the drug from their doctors and were hoarding it. The reason was that Tamiflu, as well as another potent influenza drug called Relenza (zanamivir), are viral neuraminidase inhibitors proven to be highly effective against influenza A and B viruses, including H1N1. Basically, these drugs trap the virus in the infected cell so it cannot spread to other healthy cells. This gives the body's natural immune system time to react and destroy the virus. What isn't widely known is that these drugs are effective for treatment of the common flu as well, and they can even be preventively used.

When taken after the early symptoms of the flu appear, these drugs can ameliorate the symptoms and increase the rate of recovery by up to two to three days.[8] Given this very effective response, you might think that patients would routinely be given Tamiflu or Relenza when they come down with a very severe case of the flu, but the reality is quite the opposite. It's unlikely that either of these drugs would be prescribed unless the patient was extremely sick or in a life-threatening situation. The reason for this lack of extensive use is that before Tamiflu became widely available, none of the influenza strains were resistant to it. But viruses adapt over time and since the swine flu scare, about 20 percent of flu strains have developed a resistance to Tamiflu.[9] Relenza is still very potent against most flu strains, but it is unlikely to remain so for very long. As antiviral drugs become available to the masses, viruses will inevitably build up a resistance to them. Most governments want to hold back such drugs from the mass market in order to ensure that another line of defense exists to combat more dangerous viruses when they appear.

The same is true for many antibiotics. Doctors want to avoid widespread use of these drugs so they can remain a potent weapon in their limited arsenal of options for fighting future outbreaks of virulent, life-threatening bacterial infections. On the one hand, their reasoning seems logical. After all, the risk of having a typical worker lose a week of productive time is not that expensive, but when the issue is viewed globally, the argument takes on an entirely new perspective. In many Third World nations and some European countries, prescription drugs such as Tamiflu and Relenza are readily available over the counter for a few dollars. In some Third World nations, it's even provided free at medical clinics. Thus, the attempt to forestall the gradual resistance to these drugs is futile.

Each winter these drugs are routinely abused by huge populations in developing nations, while tens of thousands of the world's most wealthy and productive scientists, researchers, and executives lose a week or more of productivity due to the common flu. It's astounding that anyone with $50 living in Turkey, Ukraine, or Russia can purchase Tamiflu while a senior executive of a major corporation in the United States is forced to spend a week in bed regardless of the level of his or her insurance coverage. On a more personal level, why should *you* lose a week or more of productivity to the flu when a readily available cure exists?

The potentially life-saving knowledge you should take away from this discussion is that just because a medication is not prescribed to you by your physician doesn't mean a medication doesn't exist to treat your condition. It's quite likely that something does exist, but is not prescribed because it's not covered by your insurance plan or still not approved in your country. The sad truth is that doctors are no longer the medical demigods they were half a century ago, dispensing ultimate wisdom with their proclamations about your health. Doctors today must walk a fine line between state-of-the-art care, what's covered by your insurance, and what might increase their risk for malpractice lawsuits. To some doctors, your care represents an insurance claim or a potential lawsuit, but to you, your care could be a matter of life and death. The best strategy is to get involved and become an informed consumer. Don't just be a passive recipient of medical care; be an active participant by staying abreast of the latest breakthroughs in medicine.

## LIFE CHOICES

Given the unsustainability of senior entitlement programs and the proclivity of politicians to only react *after* a crisis occurs, some manner of economic crisis is almost certain to occur in the future unless another major unforeseen event happens first. Unfortunately as discussed earlier, most of these unforeseen events are likely to be negative. Preparing for the uncertainties ahead will require more than savvy investing, budget balancing, and making smart health-care choices. It will also require a conscious decision to make the best possible lifestyle choices to prepare for an uncertain future.

Of course, the uninformed masses will make no changes at all. Their fate will be determined by random chance, a roll of the dice. Even many of the more informed will make the same choice to do nothing. They live in a world of denial similar to that of millions of smokers who are aware of the dangers of smoking, but choose to leave their finances, health, and their very lives to chance. Unfortunately for the smokers, it's not random chance. There is a high statistical probability that they will develop one or more serious and possibly fatal disease as a result of smoking. Similarly, there's a high probability that the coming economic crisis, regardless of how it eventually manifests itself, will create serious personal hardships.

The only realistic solution that could postpone or even eliminate the coming economic crisis would be medical advances that would postpone or eliminate aging and age-related loss of function. As you now know, these medical breakthroughs are also very likely to occur. The major uncertainty is whether they will occur in time to prevent an economic crisis due to out-of-control spending on senior entitlement programs and other welfare programs. Sadly, it's also possible that an economic crisis would occur in spite of these medical breakthroughs due to political shortsightedness. There is also a risk that by the time scientists achieve these advances, you may have already aged to the point that these new breakthroughs cannot help you. Thus, it behooves you to take proactive steps to stay as healthy as possible for as long as possible.

Managing risk requires a comprehensive approach to ensure that all major risks are addressed. A savvy homeowner wouldn't only buy fire insurance. The prudent approach would be to buy comprehensive insurance that would cover all major risks, including fire, theft, vandalism, flood, wind damage, and so on. Similarly, the prudent approach to prepare for the medical uncertainties ahead would be to take steps to protect yourself from all the major risks. Fortunately, there are some smart steps that can be taken regardless of what the future holds.

The first and most obvious step would be to take a very aggressive approach to maintaining one's own health. Since future longevity breakthroughs are likely to occur at different rates depending on the disease and a variety of issues surrounding the aging process, it makes sense to stay as healthy as possible so you can benefit from these potential breakthroughs when they eventually occur.

Your health is a very important financial asset regardless of which direction the future takes. It should be managed with the same level of personal commitment and professional oversight used to manage a $1 million stock portfolio, because over a period of two or more decades, your human capital is easily worth more than $1 million.

## GETTING INVOLVED

Regenerative medicine research is one of our best hopes to avoid the impending financial meltdown created by senior entitlement programs, yet you might be frustrated because there is so little you can do. After all, few readers of this book are likely in a position to accelerate the pace of anti-aging breakthroughs. But even if you are not a scientist, researcher, or wealthy philanthropist, you can still make a difference.

The majority of politicians and the general public are unaware of the tremendous potential benefits of regenerative medicine. They fail to grasp the profound implications that extended longevity could have on the global economy, on their respective nation's economic survival, and on their own life span and health. You can make a difference by helping to increase awareness of the exciting possibilities of regenerative medicine. Most people are still unaware that stem cells can be obtained without destroying embryos, for example. Tell your friends about iPSCs and other exciting breakthroughs just over the horizon if enough research funding can be obtained. You have a lot more influence than you might think. Posting a comment on Facebook or another social networking site might seem inconsequential, but you never know who might see and be influenced by that comment. We live in an increasingly interactive and interconnected world. Talk to your elected officials. Become politically active and support legislation that would increase funding for anti-aging research. Together, we can raise the collective awareness of the general public. Once enough people become aware of the possibilities of greatly extending their own lives, there will be a groundswell of support for regenerative medicine research.

You can also become a role model. You can influence people as much with actions as with words. It's not necessary to become fanatical about such lifestyle changes as dieting, exercise, or weight management. After only modest changes in your choices or habits, friends will begin to notice a difference. When they comment on the new and improved you, explain to them that the world is on the brink of paradigm-shifting changes, that aging itself will gradually become treatable, that major anti-aging break-throughs are likely to occur in the next couple of decades, and that you want to be healthy enough to benefit from those breakthroughs.

By New Year's Eve 2099, many of the promising breakthroughs discussed in this book will be ancient history. Some will have been surpassed by even more exotic life-extension therapies. Extreme longevity will be common throughout the developed world. Millions of healthy, active centenarians will celebrate the arrival of a new century. I plan to be one of those healthy seniors. Choose to join me. There are dark clouds on the horizon, but the distant future promises to be bright.

# Notes

## INTRODUCTION

1. G. T. McVean, L. D. Hurst, and T. Moore, "Genomic Evolution in Mice and Men: Imprinted Genes Have Little Intronic Content," *BioEssays: News and Reviews in Molecular, Cellular and Developmental Biology* 18(9) (1996): 773-75.
2. A. Bartke and H. Brown-Borg, "Life Extension in the Dwarf Mouse," *Current Topics in Developmental Biology* 63 (2004): 189-225.
3. Y. F. Chen et al., "Longevity and Lifespan Control in Mammals: Lessons from the Mouse," *Ageing Research Reviews* 9 Suppl 1 (2010): 28-35.
4. Aldous Huxley, "Case of Voluntary Ignorance," *Collected Essays* (New York: Harper, 1959).

## CHAPTER 1: APPROACHING THE TIPPING POINT

1. "2012 Annual Report of the Boards of Trustees of the Federal Hospital Insurance Trust Fund and the Federal Supplementary Medical Insurance Trust Fund," *Centers for Medicare & Medicaid Services,* Table II.B1, accessed December 2, 2012, http://www.treasury .gov/resource-center/ . . . /TR_2012_Medicare.pdf html; *The 2012 OASDI Trustees Report,* page 11, Table III.A1, IV.B2, accessed December 2, 2012, http://www.ssa.gov /oact/tr/2012/index.html.
2. Ibid. Assumes 2.9 workers/senior as shown on page 11 of the OASDI report. Other sources cited in this book find 2.1 workers/senior, which would require $12,000 in taxes per worker to pay for these programs.
3. Albert Einstein, "No amount of experimentation can . . ." at BrainyQuote, accessed November 27, 2012, http://www.brainyquote.com/quotes/quotes/a/alberteins100017 .html.
4. *Stanford Encyclopedia of Philosophy,* "Thomas Kuhn," accessed December 3, 2012, http://plato.stanford.edu/entries/thomas-kuhn/#3.
5. Max Planck quotes, at BrainyQuote, accessed November 27, 2012, http://www.brainy quote.com/quotes/authors/m/max_planck.html.
6. *Stanford Encyclopedia of Philosophy,* "Thomas Kuhn."
7. "Moore's Law," accessed November 27, 2012, http://www.mooreslaw.org/.
8. "Human Genome Project Budget," U.S. Department of Energy, accessed December 2, 2012, http://www.ornl.gov/sci/techresources/Human_Genome/project/budget.shtml.
9. N. W. S. Kam et al., "Carbon Nanotubes as Multifunctional Biological Transporters and Near-Infrared Agents for Selective Cancer Cell Destruction," *Proceedings of the National Academy of Sciences, PNAS* 102 (33) (August 16, 2005): 11600-11605.

10. "Chronic Diseases are Leading Causes of Death among Senior Citizens," *Senior Journal,* accessed November 27, 2012, http://seniorjournal.com/NEWS/SeniorStats/6-08 -07-ChronicDiseases.htm.

11. Vincent J. Cirillo, "Two Faces of Death: Fatalities From Disease And Combat In America's Principal Wars, 1775 To Present," *Perspectives in Biology and Medicine* 51 (1) (Winter 2008): 121-33, doi: 10.1353/pbm.2008.0005; Robert Marcell, "Disease in American Military History," Suite101, accessed November 28, 2012, http://suite101 .com/article/disease-in-american-military-history-a146021.

## CHAPTER 2: THE HISTORY OF LONGEVITY

1. *National Vital Statistics Reports* 59 (9) (September 28, 2011): Table 20.

2. *Encyclopedia of Life,* "Facts about Humans (*Homo sapiens*)," accessed November 28, 2012, http://eol.org/pages/327955/details#cite_note-5.

3. *Encyclopedia of Ancient Egypt,* "Maps, Timeline, Information about the Dynasties, Pharaohs, Laws, Culture, Government, Military and More," Google Books, accessed November 28, 2012, http://books.google.com/books?id=MwvM09Z-7DwC &source=gbs_navlinks_s; Ian Shaw, ed. *The Oxford History of Ancient Egypt* (Oxford: Oxford University Press, 2000).

4. Ibid.

5. Ibid.

6. Ibid.

7. "Middle Pleistocene Lower Back and Pelvis from an Aged Human Individual from the Sima de los Huesos Site, Spain," *Proceedings of the National Academy of Sciences USA,* 107(43) (October 26, 2010): 18386-18391, accessed November 28, 2012, http://www .ncbi.nlm.nih.gov/pmc/articles/PMC2973007/.

8. "Dr. Semmelweis' Biography," Semmelweis Society International, accessed November 28, 2012, http://semmelweis.org/about/dr-semmelweis-biography/.

9. Ibid.

10. "Handwashing," Access Excellence, accessed November 28, 2012, http://www.access excellence.org/AE/AEC/CC/hand_background.php.

11. "History of Penicillin–Alexander Fleming–John Sheehan–Andrew Moyer," About .com, Inventors, accessed November 28, 2012, http://inventors.about.com/od/pstart inventions/a/Penicillin.htm.

12. Ibid.

13. Ibid.

14. Ibid.

15. Ibid.

16. Tom Kirkwood, "The End of Age," *Journal of Advanced Nursing* (London: BBC in association with Profile books, 2001), accessed November 28, 2012, http://onlinelibrary .wiley.com/doi/10.1046/j.1365-2648.2001.2045b.x/abstract.

17. "World Economic Outlook Database October 2012," accessed November 28, 2012, http://www.imf.org/external/pubs/ft/weo/2012/02/weodata/index.aspx; "United Nations World Population Prospects: 2006 revision—Table A.17 for 2005-2010," accessed November 28, 2012, http://www.un.org/esa/population/publications/wpp2006 /WPP2006_Highlights_rev.pdf.

18. Ibid.

19. "Health, United States, 2010, Table 22. Life Expectancy at Birth, at 65 Years of Age, and at 75 Years of Age, by Race and Sex: United States, Selected Years 1900-2007," Centers for Disease Control and Prevention, accessed November 28, 2012, http://www .cdc.gov/nchs/data/hus/2010/022.pdf.

20. "Papers of George Washington," George Washington Papers, accessed November 28, 2012, http://gwpapers.virginia.edu/articles/wallenborn.html.
21. "Youngtown—The First Active Retirement Community," 55 Places, accessed November 28, 2012, http://www.55places.com/blog/youngtown-active-retirement-community.
22. John Findlay, *Magic Lands: Western Cityscapes and American Culture After 1940* (Berkeley: University of California Press, 1992).
23. "About Del Webb," Del Webb, accessed December 3, 2012, http://www.delwebb.com/value-of-delwebb/About_DelWebb.aspx.
24. Findlay, *Magic Lands.*
25. "The History of Sun City," Recreation Centers of Sun City, http://www.sunaz.com/history.
26. Findlay, *Magic Lands.*
27. Ibid.
28. Ibid.
29. *The 2012 OASDI Trustees Report,* Social Security: The Official Website of the U.S. Social Security Administration, accessed December 2, 2012, http://www.ssa.gov/oact/tr/2012/index.html.
30. Grady Cash, personal interview, November 11, 2011.

## CHAPTER 3: DEMOGRAPHICS OF MAJOR ECONOMIES: THE UNITED STATES, THE EUROPEAN UNION, RUSSIA, CHINA, AND JAPAN

1. "World Fertility Policies 2011," United Nations, Department of Economic and Social Affairs, Population Division, accessed November 28, 2012, http://www.un.org/esa/population/publications/worldfertilitypolicies2011/wfpolicies2011.html.
2. "Population–The 2012 Statistical Abstract," U.S. Census Bureau, accessed November 28, 2012, http://www.census.gov/compendia/statab/cats/population.html.
3. "1992 Presidential General Election Results," U.S. Election Atlas, accessed November 28, 2012, http://uselectionatlas.org/RESULTS/national.php?year=1992.
4. John Nichols, "Young Voter Turnout Fell 60% from 2008 to 2010; Dems Won't Win in 2012 If the Trend Continues," *The Nation,* November 16, 2010, accessed November 28, 2012, http://www.thenation.com/blog/156470/young-voter-turnout-fell-60-2008-2010-dems-wont-win-2012-if-trend-continues#.
5. David Paul Kuhn, "The Senior Wave: Older Voters Set for Historic Turnout," RealClearPolitics, October 18, 2010, accessed November 28, 2012, http://www.realclearpolitics.com/articles/2010/10/18/the_senior_wave_older_voters_set_for_historic_turnout_107608.html.
6. Ted C. Fishman, *Shock of Gray* (New York: Scribner, 2010), 163.
7. X. Bai et al., "Evaluation of Biological Aging Process—A Population-based Study of Healthy People in China," *Gerontology* 56(2) (2010): 129-40.
8. P. Arnsberger, P. Fox, X. Zhang, and S. Gui, "Population Aging and the Need for Long Term Care: A Comparison of the United States and the People's Republic of China," *Journal of Cross-Cultural Gerontology* 15(3) (2000): 207-27.
9. Bai et al., "Evaluation of Biological Aging Process."
10. P. Du, "Fertility Decline and Population Aging in China," *Chinese Journal of Population Science* 7(4) (1995): 299-306.
11. "World Population Ageing 2009," United Nations, Department of Economic and Social Affairs Population Division, December 2009, accessed November 28, 2012, http://www.un.org/esa/population/ . . . /WPA2009_WorkingPaper.pdf.

12. Elise Young, "China and India Producing Larger Share of Global College Graduates," *Inside Higher Ed,* July 12, 2012, accessed November 29, 2012, http://www.inside highered.com/news/2012/07/12/china-and-india-producing-larger-share-global -college-graduates.

13. Joseph Casey and Katherine Koleski, *Backgrounder: China's 12th Five-Year Plan,* U.S.-China Economic & Security Review Commission, June 24, 2011, 5-10, http://www .uscc.gov/researchpapers/2011/12th-FiveYearPlan_062811.pdf.

14. J. Bijak et al., "Population and Labour Force Projections for 27 European Countries, 2002-2052: Impact of International Migration on Population Ageing," *European Journal of Population* 23(1) (2007): 1-31.

15. Matthew Karnitschig, "German Immigration Policy: Merkel Enters Policy Fray," *Wall Street Journal,* October 18, 2010, accessed November 29, 2012, http://online.wsj.com /article/SB10001424052702304250404575558583224907168.html.

16. "Auto Accidents Kill Some 28,000 Russians in 2011," *RIA Novosti,* February 2, 2012, accessed November 29, 2012, http://en.rian.ru/russia/20120210/171250033 .html; "Traffic Deaths In 2011 Fell To Record Low In U.S," Huffington Post, May 2, 2012, http://www.huffingtonpost.com/2012/05/08/us-2011-traffic-deaths-record-low _n_1498344.html.

17. "Europe and Central Asia–ECA KNOWLEDGE BRIEF: Dangerous Roads–Russia's Safety Challenge," World Bank, accessed November 29, 2012, http://web.worldbank .org/WBSITE/EXTERNAL/COUNTRIES/ECAEXT/0,contentMDK:22659724 ~pagePK:146736~piPK:146830~theSitePK:258599~isCURL:Y,00.html.

18. Michael Erikson, *The Tobacco Atlas,* 4th Edition (New York: American Cancer Society, 2012), 132.

19. "Homicide Statistics 2012," United Nations Office on Drugs and Crime, accessed November 29, 2012, http://www.unodc.org/documents/data-and-analysis/statistics/crime /Homicide_statistics2012.xls.

20. "The Future of AIDS–Society and Culture,"AEI, accessed November 29, 2012, http:// www.aei.org/article/society-and-culture/citizenship/the-future-of-aids/.

21. Ibid.

22. "International Programs–Information Gateway," U.S. Census Bureau, accessed November 29, 2012, http://www.census.gov/population/international/data/idb/information Gateway.php.

23. Ibid.

24. "Russia–Natural Resources," Country Studies, accessed November 29, 2012, http:// countrystudies.us/russia/59.htm.

25. Paul Ehrlich, *The Population Bomb* (New York: Ballantine, 1968).

26. *World Population—United States* (82): 2, Bureau of the Census, International Statistical Programs Center (U.S.), Google Books, accessed December 3, 2012, http://books .google.com/books?id=RlIvAAAAMAAJ&pg=PA2&dq=world+population+1968&hl =en&sa=X&ei=Bsm8ULLjNJCC9gS_joGwBA&ved=0CDwQ6AEwAg#v=onepage &q=world%20population%201968&f=false.

27. Fairfield Osborn, *Our Plundered Planet* (Boston: Little, Brown, and Company, 1948), 17.

## CHAPTER 4: AGING AND LOSS OF FUNCTION

1. "Hearing Loss," Johns Hopkins Medicine Health Library, accessed November 29, 2012, http://www.hopkinsmedicine.org/healthlibrary/conditions/otolaryngology/hearing _loss_85,P00451/.

2.  "Aging Changes In the Senses," *MedlinePlus Medical Encyclopedia,* accessed November 29, 2012, http://www.nlm.nih.gov/medlineplus/ency/article/004013.htm.

3.  A. Mackay-Sim et al., "Olfactory Ability In the Healthy Population: Reassessing Presbyosmia," *Chem Senses,* October 31, 2006 (8):763-71, accessed November 29, 2012, http://www.ncbi.nlm.nih.gov/pubmed/16901951.

4.  Ming Wang, "Primary Eye Care in China—An Emerging Field with Increasing Opportunities," Tennessee Chinese Chamber of Commerce, accessed November 29, 2012, http://www.tccc.us/articles/40-primary-eye-care-in-china-an-emerging-field-with -increasing-opportunities.html.

5.  J. S. Schiller et al., "Summary Health Statistics for U.S. Adults: National Health Interview Survey, 2010," *Vital and Health Statistics Series 10, Data from the National Health Survey* 252 (2012): 1-207.

6.  W. W. Campbell et al., "Increased Energy Requirements and Changes in Body Composition with Resistance Training in Older Adults," *The American Journal of Clinical Nutrition* 60(2) (1994): 167-75.

7.  Elizabeth L. Glisky, "Changes in Cognitive Function in Human Aging–Brain Aging," in *Brain Aging: Models, Methods, and Mechanisms,* ed. D.R. Riddle (Boca Raton, FL: CRC Press, 2007), accessed November 29, 2012, http://www.ncbi.nlm.nih.gov/books /NBK3885/.

8.  R. Brookmeyer et al., "National Estimates of the Prevalence of Alzheimer's Disease in the United States," *Alzheimer's & Dementia: The Journal of the Alzheimer's Association* 7(1) (2011): 61-73.

9.  R. Helson, "Personality Change Over 40 Years of Adulthood," *Journal of Personality and Social Psychology* (2002), accessed November 29, 2012, http://www.ncbi.nlm.nih.gov /pubmed/12219867.

10.  K. Warner Schaie, "The Seattle Longitudinal Study: Relationship Between Personality and Cognition," accessed November 29, 2012, http://www.ncbi.nlm.nih.gov/pmc /articles/PMC1474018/.

11.  Patricia A. Reuter-Lorenz et al., "Age Differences in the Frontal Lateralization of Verbal and Spatial Working Memory Revealed by PET," *Journal of Cognitive Neuroscience* 12(1): 174-87, accessed November 29, 2012, http://www-personal.umich.edu/~jjonides /pdf/2000_3.pdf.

12.  "The young man knows the rules . . ." at BrainyQuote, accessed November 29, 2012, http://www.brainyquote.com/quotes/quotes/o/oliverwend108494.html.

13.  T. A. Salthouse, *Cognitive Aging: A Primer* (New York: Psychology Press, 2000).

14.  L. B. Goldstein et al., "Guidelines for the Primary Prevention of Stroke: A Guideline for Healthcare Professionals from the American Heart Association/American Stroke Association," *Stroke: A Journal of Cerebral Circulation* 42(2) (2011): 517-84.

## CHAPTER 5: BIOLOGIC AGING AT A GLANCE

1.  "Testosterone Side Effects," Drugs.com, accessed November 30, 2012, http://www .drugs.com/sfx/testosterone-side-effects.html.

2.  Michael R. Rose, "The Evolution of Animal Senescence," *Canadian Journal of Zoology* 62(9) (1984): 1661-67, doi: 10.1139/z84-243.

3.  "Cellular Senescence: The Hayflick Limit and Aging Cells," senescence.info, accessed November 30, 2012, http://www.senescence.info/cell_aging.html.

4.  "Rate-of-Living," Pysch527 course page, Cornell University, accessed November 30, 2012, https://courses.cit.cornell.edu/psych527_nbb420-720/student2005/nrb26/Page _2.htm.

5.  "Hummingbird Facts," World of Hummingbirds, accessed November 30, 2012, http://www.worldofhummingbirds.com/facts.php; "Adult Elephant Life Cycle," Elephant Information Repository, accessed November 30, 2012, http://elephant.elehost.com/About_Elephants/Life_Cycles/Adult/adult.html.

6.  C. Rocken, W. Saeger, and R. P. Linke, "Gastrointestinal Amyloid Deposits in Old Age: Report on 110 Consecutive Autopsical Patients and 98 Retrospective Bioptic Specimens," *Pathology, Research and Practice,* 190(7) (August 1994): 641, accessed November 30, 2012, http://www.ncbi.nlm.nih.gov/pubmed/7808962.

## CHAPTER 6: REPAIRING DAMAGE AND EXTENDING WORK SPAN

1.  Average Social Security payments are about $14,000 annually. Each five-year age group of baby boomers is about 21 million people, of whom 62 percent will draw Social Security checks. $14,000 x 21,000,000 x 0.62 = $182 million.

2.  "The 'Silver Tsunami': Why Older Workers Offer Better Value Than Younger Ones," Knowledge@Wharton, December 6, 2010, accessed November 30, 2012, http://knowledge.wharton.upenn.edu/article.cfm?articleid=2644.

3.  "Why Employers Don't Want to Hire Boomers," Executive Coaching and Job Search Coaching, accessed November 30, 2012, http://jobsearch4execs.com/2010/11/17/why-employers-dont-want-to-hire-boomers/.

4.  Ibid.

5.  "Older Workers, Costs of," Sloan Center for Aging and Work, accessed November 30, 2012, http://capricorn.bc.edu/agingandwork/database/browse/facts/fact_record/5663/all.

6.  "Why Employers Don't Want to Hire Boomers," Executive Coaching and Job Search Coaching.

7.  Ibid.

8.  William Safire, "Language: Tracking the Source of the 'Third Rail' Warning," *International Herald Tribune–The New York Times,* February 18, 2007, accessed November 30, 2012, http://www.nytimes.com/2007/02/18/opinion/18iht-edsafmon.4632394.html.

9.  K. M. Flegal et al., "Prevalence of Obesity and Trends in the Distribution of Body Mass Index among U.S. Adults, 1999-2010," *JAMA: The Journal of the American Medical Association* 307(5) (2012): 491-97.

10. K. D. Bertakis and R. Azari, "Obesity and the Use of Health Care Services," *Obesity Research* 13(2) (2005): 372-79.

11. "Diabetes and Obesity Increase Risk for Breast Cancer Development," EurekAlert!, accessed December 5, 2012, http://www.eurekalert.org/pub_releases/2011-12/aafc-dao120111.php.

12. M. Kivipelto et al., "Obesity and Vascular Risk Factors at Midlife and the Risk of Dementia and Alzheimer Disease," *Arch Neurol* 62(10) (October 2005): 1556-60.

## CHAPTER 7: RECENT ADVANCES IN BIOGERONTOLOGY AND REGENERATIVE MEDICINE

1.  MeSH, "Regenerative Medicine," accessed November 30, 2012, http://www.ncbi.nlm.nih.gov/mesh/68044968.

2.  Sheryl Stolberg, "Obama Is Leaving Some Stem Cell Issues to Congress," *New York Times,* March 9, 2009, accessed November 30, 2012, http://www.nytimes.com/2009/03/09/us/politics/09stem.html.

3.  "Wisconsin Institutes for Discovery—James Thomson Home," Wisconsin Institutes for Discovery, accessed November 30, 2012, http://discovery.wisc.edu/home/morgridge/about-morgridge/leadership/james-thomson/james-thomson-home.cmsx.

4. Sally Lehrman, "Dolly's Creator Moves Away from Cloning and Embryonic Stem Cells," *Scientific American,* August 2008, accessed December 1, 2012, http://www.scientificamerican.com/article.cfm?id=no-more-cloning-around.

5. "Shinya Yamanaka Wins 2012 Nobel Prize in Medicine," University of California San Francisco website, accessed December 1, 2012, http://www.ucsf.edu/news/2012/10/12898/shinya-yamanaka-wins-2012-nobel-prize-medicine.

6. "'Universal' Virus-Free Method Developed By Scientists to Turn Blood Cells into Beating Heart Cells," Medical News Today, accessed December 1, 2012, http://www.medicalnewstoday.com/releases/221898.php.

7. Ibid.

8. P. Huang et al., "Induction of Functional Hepatocyte-like Cells from Mouse Fibroblasts by Defined Factors," *Nature* 475(7356) (May 11, 2011): 386-9, doi: 10.1038/nature10116.

9. Heidi Ledford, "Reprogrammed Cells Repair Damaged Livers," *Nature News,* accessed December 1, 2012, http://www.nature.com/news/2011/110511/full/news.2011.283.html.

10. "The Gift of a Lifetime—History of Transplantation," OrganTransplants.org, accessed December 1, 2012, http://www.organtransplants.org/understanding/history/index.html.

11. Ibid.

12. Ibid.

13. United Network for Organ Sharing, accessed November 30, 2012, http://www.unos.org/. This site updates daily. Data as of November 30, 2012, showed 116,727 on the waiting list and 18,986 transplants in the first eight months of 2012.

14. J. Stehlik et al., "The Registry of the International Society for Heart and Lung Transplantation: Twenty-seventh Official Adult Heart Transplant Report—2010," *The Journal of Heart and Lung Transplantation: The Official Publication of the International Society for Heart Transplantation* 29(10) (2010): 1089-103.

15. S.L. Murphy, J. Q. Xu, and K. D. Kochanek, "Deaths: Preliminary Data for 2010," *National Vital Statistics Reports* 60(4): Table 2, accessed December 1, 2012, www.cdc.gov/nchs/data/nvsr/nvsr60/nvsr60_04.pdf.

16. "Jarvik Heart Resources—Robert Jarvik on the Jarvik-7,"accessed December 1, 2012, http://www.jarvikheart.com/basic.asp?id=69.

17. "William Schroeder Dies 620 Days After Receiving Artificial Heart," *New York Times,* August 7, 1986.

18. S. Westaby et al., "First Permanent Implant of the Jarvik 2000 Heart," *Lancet* 356(9233) (2000): 900-903.

19. A. Meyer and M. Slaughter, "The Total Artificial Heart," *Panminerva Medica* 53(3) (2011): 141-54; "The Total Artificial Heart," NextBio, accessed December 1, 2012, http://www.nextbio.com/b/search/article.nb?id=21775941.

20. "VAD Product Information," Thoratec CentriMag, Thoratec Corporation, accessed December 1, 2012, http://www.thoratec.com/medical-professionals/vad-product-information/thoratec-centrimag.aspx.

21. Ibid.

22. O. H. Frazier et al., "Optimization of Axial-pump Pressure Sensitivity for a Continuous-flow Total Artificial Heart," *The Journal of Heart and Lung Transplantation: The Official Publication of the International Society for Heart Transplantation* 29(6) (2010): 687-91.

23. B. Levi et al., "Nonintegrating Knockdown and Customized Scaffold Design Enhances Human Adipose-Derived Stem Cells in Skeletal Repair," *Stem Cells,* 29(12) (2011): 2018-29.

24. "Anthony Atala at TEDMED 2009," YouTube video clip, accessed December 1, 2012, http://www.youtube.com/watch?v=QQP3HrU8fdM.

25. Mikhail S. Shchepinov, "Reactive Oxygen Species, Isotope Effect, Essential Nutrients, and Enhanced Longevity," *Rejuvenation Research* 10(1) (March 1, 2007): 47-60. doi:10.1089/rej.2006.0506. PMID 17378752.

26. S. J. Grabowski et al., "Quantitative Classification of Covalent and Noncovalent H-bonds," *The Journal of Physical Chemistry B* 110(13) (2006): 6444-46.

27. D. A. Knorr and D. S. Kaufman, "Pluripotent Stem Cell–derived Natural Killer Cells for Cancer Therapy," *Translational Research: The Journal of Laboratory and Clinical Medicine* 156(3) (2010): 147-54.

28. A. Moskalev et al., "The Role of D-GADD45 in Oxidative, Thermal and Genotoxic Stress Resistance," *Cell Cycle* 11(22) (October 24, 2012).

29. "Stem Cell Research Promising," Xinhua.net, accessed December 2, 2012, http://news.xinhuanet.com/english2010/china/2011-01/29/c_13712526.htm.

30. Joshua Berlin, "With Government Support, China Medical City Looks to Separate Itself from the Pack," *PharmAsia News,* August 13, 2011, accessed December 1, 2012, www.cctec.cornell.edu/news/PharmAsiaNews-082011.pdf.

31. "World Fertility Policies 2011," United Nations, Department of Economic and Social Affairs, Population Division, accessed November 28, 2012, http://www.un.org/esa/population/publications/worldfertilitypolicies2011/wfpolicies2011.html

## CHAPTER 8: THE REAL COSTS OF AGING

1. "Funding Savings Needed for Health Expenses for Persons Eligible for Medicare," Employee Benefit Research Institute, accessed December 2, 2012, http://www.ebri.org/publications/ib/index.cfm?fa=ibDisp&content_id=4711.

2. P. Hemp, "Presenteeism: At Work—But Out of It," *Harvard Business Review* 82(10) (2004): 49-58, 155.

3. "The 2012 OASDI Trustees Report," United States Social Security Administration, accessed December 2, 2012, http://www.ssa.gov/oact/tr/2012/index.html.

4. Ibid. Table II.B1.

5. "2012 Annual Report of the Boards of Trustees of the Federal Hospital Insurance Trust Fund and the Federal Supplementary Medical Insurance Trust Fund," Centers for Medicare & Medicaid Services, accessed December 2, 2012, http://www.treasury.gov/resource-center/ . . . /TR_2012_Medicare.pdf html.

6. "Pensions at a Glance 2011,"*OECD Pensions Statistics,* doi: 10.1787/data-00625-en.

7. M. L. Daviglus et al., "Relationship of Fruit and Vegetable Consumption in Middle-aged Men to Medicare Expenditures in Older Age: The Chicago Western Electric Study," *Journal of the American Dietetic Association* 105(11) (2005): 1735-44.

8. K. M. Flegal et al., "Prevalence and Trends in Obesity among U.S. Adults, 1999-2008," *JAMA: The Journal of the American Medical Association* 303(3) (2010): 235-41.

9. A. Sansone-Parsons et al., "Effects of Age, Gender, and Race/Ethnicity on the Pharmacokinetics of Posaconazole in Healthy Volunteers," *Antimicrobial Agents and Chemotherapy* 51(2) (2007): 495-502.

10. "CMS Announces Medicare Premiums, Deductibles, and Coinsurance Rates for 2013," *CCH, Wolters Kluwer Law & Business*, accessed December 5, 2012, http://hr.cch.com/news/uiss/120412a.asp.

11. "2012 Annual Report of the Boards of Trustees of the Federal Hospital Insurance Trust . . ." Centers for Medicare & Medicaid Services.

12. $6 trillion divided by 130 million workers equals $46,153 per worker.

13. "AssetBuilder–How Much Can the 99% Squeeze Out of the 1%?" AssetBuilder Inc., October 21, 2011, accessed December 2, 2012, http://assetbuilder.com/blogs/scott _burns/archive/2011/10/21/how-much-can-the-99-squeeze-out-of-the-1.aspx.

14. William A. McEachern, *Contemporary Economics* (Hampshire: South-Western Educational Publishing, 2006), 530.

## CHAPTER 9: CHANGING THE PRIORITIES OF MEDICAL RESEARCH

1. A. Jemal et al., "Cancer Statistics, 2010," *CA: A Cancer Journal for Clinicians* 60(5) (2010): 277-300.

2. S. R. Hinchliffe, P. W. Dickman, and P. C. Lambert, "Adjusting for the Proportion of Cancer Deaths in the General Population when Using Relative Survival: A Sensitivity Analysis," *Cancer Epidemiology* 36(2) (2012): 148-52.

3. Figures obtained by mining the International Aging Research Portfolio (IARP) database and FundingTrends.org.

4. "Mission—About NIH," National Institutes of Health, accessed December 2, 2012, http://www.nih.gov/about/mission.htm.

5. R. G. Victor et al., "A Barber-based Intervention for Hypertension in African American Men: Design of a Group Randomized Trial," *American Heart Journal* 157(1) (2009): 30-36.

6. These researchers obviously felt their project had merit, so in order not to embarrass these individuals unnecessarily, this research project has been modified slightly to prevent its identification. Readers can search a list of research projects using the NIH Reporter search engine at www.reporter.nih.gov and decide for themselves the value of some current research projects.

7. "Big Pharma's Political Contributions," *The Washington Post,* January 8, 2009, accessed December 2, 2012, http://www.washingtonpost.com/wp-dyn/content/graphic /2009/01/08/GR2009010800559.html.

8. D. Jiang, "Myths about Doing Business in China: This Large and Growing Country Provides Many Opportunities for Companies in the Know," *Genetic Engineering & Biotechnology News* 32(5) (2012): 21-22.

9. "Annual Report of the Boards of Trustees of the Federal Hospital Insurance and Federal Supplemental Medical Insurance: Big Pharma's Political Contributions Trust Funds," Centers for Medicare & Medicaid Services, 2011, http://www.cms.gov%2FResearch -Statistics-Data-and-Systems%2FStatistics-Trends-and-Reports%2FReportsTrust Funds%2Fdownloads%2Ftr2011.pdf.

## PART IV: THE RETIREMENT CULTURE

1. A. Zhavoronkov, N. Mirza, I. Artyuhov, and E. Debonneuil, "Evaluating the Impact of Recent Advances in Biomedical Sciences and the Possible Mortality Decreases on the Future of Health Care and Social Security in the United States," *Pensions International Journal* (2012): 1-11.

## CHAPTER 10: CHANGING THE RETIREMENT CULTURE

1. D. J.Ekerdt, "Born to Retire: The Foreshortened Life Course," *The Gerontologist* 44(1) (2004): 3-9.

2. Ibid.

3. "New Places to Live When You Retire," *The Kiplinger Magazine,* October 1965, 25.

4.  "History," KFC.com, accessed December 6, 2012, http://www.kfc.com/about/history.asp.

5.  "Cornelius Vanderbilt," Articles, Video, Pictures and Facts, History.com, accessed December 6, 2012, http://www.history.com/topics/cornelius-vanderbilt.

6.  Ibid.

7.  "Mary Kay Ash Biography—Facts, Birthday, Life Story," Biography.com, accessed December 6, 2012, http://www.biography.com/people/mary-kay-ash-197044.

8.  "Mary Kay Ash—Most Outstanding Woman in Business in the 20th Century," About.com, Entrepreneurs, accessed December 6, 2012, http://entrepreneurs.about.com/od/famousentrepreneurs/p/marykayash.htm.

9.  "John Jacob Astor Is Born," History.com, This Day in History—1/17/1763, accessed December 6, 2012, http://www.history.com/this-day-in-history/john-jacob-astor-is-born.

10. Ibid.

11. Tripti Lahiri, "Specs for India's $35 Aakash Tablet–India Real Time," *Wall Street Journal,* October 25, 2011, accessed December 5, 2012, http://blogs.wsj.com/indiarealtime/2011/10/05/specs-of-indias-35-tablet.

12. "Jobless Claims in U.S. Decrease as Sandy Effect Dissipates," *Businessweek,* November 29, 2012, accessed December 2, 2012, http://www.businessweek.com/news/2012-11-29/initial-jobless-claims-in-u-dot-s-dot-fall-as-sandy-effect-dissipates. The "over 5 million" figure reflects 3.29 million unemployment insurance recipients plus 2.16 million receiving extended unemployment benefits. The data in this chapter is shown as "over 5 million" rather than a specific figure since the numbers are constantly changing.

13. "University Catalog," Coursera, accessed December 6, 2012, https://www.coursera.org/universities.

14. "Completely Free Online Classes? Coursera.org Now Offering Courses from 16 Top Colleges," TED Blog, accessed December 2, 2012, http://blog.ted.com/2012/07/18/completely-free-online-classes-coursera-org-now-offering-courses-from-16-top-colleges/.

15. "Coursera Reaches 1 Million Students Worldwide—Tech News and Analysis," GigaOM, accessed December 2, 2012, http://gigaom.com/2012/08/09/coursera-reaches-1-million-students-worldwide.

## CHAPTER 11: NATURE VERSUS NURTURE: REVERSING PSYCHOLOGICAL AGING

1.  Alexandra M. Freund and Paul B. Baltes, "Selection, Optimization, and Compensation as Strategies of Life Management: Correlations with Subjective Indicators of Successful Aging." *Psychology and Aging* 13(4) (December 1998): 531-43.

## CHAPTER 12: PREVENTIVE AND REGENERATIVE MEDICINE

1.  "The Burden of Chronic Disease on Business and U.S. Competitiveness," *2009 Almanac of Chronic Disease* (Partnership to Fight Chronic Disease, 2009), 35-50, accessed December 2, 2012, http://www.prevent.org/data/files/News/pfcdalmanac_excerpt.pdf.

2.  "Wellness Guide to Preventive Care," Wellness Letter, accessed December 2, 2012, http://www.wellnessletter.com/ucberkeley/foundations/preventive-care/#.

3.  R. A. Smith, V. Cokkinides, and H. J. Eyre, "American Cancer Society Guidelines for the Early Detection of Cancer, 2006," *CA: A Cancer Journal for Clinicians* 56(1) (2006): 11-25; quiz 49-50.

4.  K. Thanapirom, S. Treeprasertsuk, and R. Rerknimitr, "Awareness of Colorectal Cancer Screening in Primary Care Physicians," *Journal of the Medical Association of Thailand Chotmaihet thangphaet* 95(7) (2012): 859-65.

5. "Diabetes Statistics," American Diabetes Association, accessed December 2, 2012, http://www.diabetes.org/diabetes-basics/diabetes-statistics/. Note that this web page is periodically updated. Data shown is that appearing on the page on the access date.

6. T. Yates et al., "The Pre-diabetes Risk Education and Physical Activity Recommendation and Encouragement (PREPARE) Programme Study: Are Improvements in Glucose Regulation Sustained at 2 Years?" *Diabetic Medicine: A Journal of the British Diabetic Association* 28(10) (2011): 1268-71.

7. W. C. Nichols et al., "Genetic Screening for a Single Common LRRK2 Mutation in Familial Parkinson's Disease," *Lancet* 365(9457) (2005): 410-12.

8. E. Sidransky et al., "Multicenter Analysis of Glucocerebrosidase Mutations in Parkinson's Disease," *The New England Journal of Medicine* 361(17) (2009): 1651-61.

9. "Human Genome Project Budget," Department of Energy, accessed December 2, 2012, http://www.ornl.gov/sci/techresources/Human_Genome/project/budget.shtml.

10. S. T. Bennett et al.,"Toward the 1,000 Dollars Human Genome," *Pharmacogenomics* 6(4) (2005): 373-82.

11. "Life Technologies (LIFE): Ion World 2012 A Potential Catalyst," iStockAnalyst.com, accessed December 2, 2012, http://www.istockanalyst.com/finance/story/6021664/life-technologies-life-ion-world-2012-a-potential-catalyst.

12. Benjamin Zycher, "Medical Progress Report 5—Comparing Public and Private Health Insurance," Manhattan Institute for Policy Research, October 2007, accessed December 2, 2012, http://www.manhattan-institute.org/html/mpr_05.htm.

13. "The National Fiscal Year 2012," National Institutes of Health, accessed December 2, 2012, http://officeofbudget.od.nih.gov/cy.html.

14. Parija Kavilanz, "RX for Money Woes: Doctors Want to Call It Quits Because of the Costs," CNN Blogs, September 14, 2009, accessed December 2, 2012, http://money.cnn.com/2009/09/14/news/economy/health_care_doctors_quitting/index.htm.

15. Parija Kavilanz, "Health Reform Sparks Interest In Primary Care," CNN Blogs, March 23, 2011, accessed December 2, 2012, http://money.cnn.com/2011/03/23/news/economy/more_medical_students_pick_family_medicine/index.htm.

16. Peggy Peck, "Malpractice Adds Less than 3 Percent to Healthcare Tab," *MedPage Today*, September 7, 2010, accessed December 2, 2012, http://www.medpagetoday.com/PracticeManagement/Medicolegal/22051.

17. P. E. Cramer et al., "ApoE-directed Therapeutics Rapidly Clear Beta-amyloid and Reverse Deficits in AD Mouse Models," *Science* 335(6075) (2012): 1503-6.

18. Ibid.

19. "2012 Alzheimer's Disease Facts and Figures," *Alzheimer's & Dementia* 8(2), Alzheimer's Association, accessed December 2, 2012, http://www.alz.org/downloads/facts_figures_2012.pdf.

20. R. A. Hammond and R. Levine, "The Economic Impact of Obesity In the United States," *Diabetes, Metabolic Syndrome and Obesity: Targets and Therapy* 3 (2010): 285-95.

21. S. Calfo et al, Centers for Medicare & Medicaid Services, Office of the Actuary Report, "Last Year of Life Study," 2008, http://www.cms.gov/Research-Statistics-Data-and-Systems/Research/ActuarialStudies/downloads/Last_Year_of_Life.pdf.

22. Ibid.

23. N.G. Levinsky et al., "Influence of Age on Medicare Expenditures and Medical Care in the Last Year of Life," *JAMA: The Journal of the American Medical Association* 86(11) (2001): 1349-55.

24. Ken Murray, "How Doctors Die," Zocalo Public Square, accessed December 2, 2012, http://www.zocalopublicsquare.org/2011/11/30/how-doctors-die/ideas/nexus/.

25. Grady Cash, personal interview, November 11, 2011.

26. Cathi Hanauer, "Buying Time: *The Cost of Hope* by Amanda Bennett," *New York Times* July 22, 2012, accessed December 2, 2012, http://www.nytimes.com/2012/07/22 /books/review/the-cost-of-hope-by-amanda-bennett.html.

27. S. Calfo et al, Centers for Medicare & Medicaid Services, Office of the Actuary Report, "Last Year of Life Study."

28. Michael Bowen, "ICU Expenses," Cobb, June 2, 2010, accessed December 2, 2012, http://cobb.typepad.com/cobb/2010/06/icu-expenses.html.

## CHAPTER 13: PREPARING FOR THE FUTURE

1. "Funding Savings Needed for Health Expenses for Persons Eligible for Medicare," *EBRI Issue Brief* 351 (December 2010), accessed December 2, 2012, http://www.ebri.org /publications/ib/index.cfm?fa=ibDisp&content_id=471.

2. Ibid.

3. "2013 Federal Income Tax Brackets And Marginal Rates," *Forbes,* accessed December 5, 2012, http://www.forbes.com/sites/moneybuilder/2012/11/01/2013-federal-income -tax-brackets-and-marginal-rates/. Note that tax brackets are subject to change frequently. While future rates may be different, the principle of this example still applies.

4. "Exercise," Thomas Jefferson's Monticello, accessed December 5, 2012, http://www .monticello.org/site/research-and-collections/exercise.

5. J. C. LaRosa, J. He, and S. Vupputuri, "Effect of Statins on Risk of Coronary Disease: A Meta-analysis of Randomized Controlled Trials," *JAMA: The Journal of the American Medical Association* 282(24) (1999): 2340-46.

6. K. M. Dale et al., "Statins and Cancer Risk: A Meta-analysis," *JAMA: The Journal of the American Medical Association* 295(1) (2006): 74-80.

7. A. Agarwal, "Do Companion Diagnostics Make Economic Sense for Drug Developers?" *New Biotechnology* 29(6) (2012): 695-708.

8. R. Dutkowski, "Oseltamivir in Seasonal Influenza: Cumulative Experience in Low- and High-risk Patients," *The Journal of Antimicrobial Chemotherapy* 65 Suppl 2 (2010): 11-24.

9. M. Lipsitch et al., "Antiviral Resistance and the Control of Pandemic Influenza," *PLoS Medicine* 4(1) (2007): 15.

# Index